Biophysical Aerodynamics and the Natural Environment

Biophysical Aerodynamics and the Natural Environment

A. J. Ward-Smith
Department of Mechanical Engineering
Brunel University

A Wiley–Interscience Publication

JOHN WILEY & SONS
Chichester . New York . Brisbane . Toronto . Singapore

Library of Congress Cataloging in Publication Data:

Ward-Smith, A. J. (Alfred John)
Biophysical aerodynamics and the natural environment.
'A Wiley–Interscience publication.'
Bibliography: p.
Includes index.
1. Animal flight. 2. Biophysics. 3. Aerodynamics.
I. Title.
QP310.F5W37 1984 591.1′852 84-11870

ISBN 0 471 90436 8

British Library Cataloguing in Publication Data:

Ward-Smith, A. J.
Biophysical aerodynamics and the natural
environment.
1. Biophysics 2. Aerodynamics
I. Title
574.19′1 QH505
ISBN 0 471 90436 8

Typeset by Activity Limited, Salisbury, Wilts, and printed
St Edmundsbury Press, Bury St Edmunds, Suffolk

To
Barbara
and to
Sharon and Wayland

Contents

Preface

Chapter 1. General Introduction to Aerodynamic Principles **1**
1.1 Physical Properties of Air and the Atmosphere 2
1.2 Relative Motion 7
1.3 Reynolds Number 8
1.4 The Flow Field about a Moving Body 10
1.5 Aerodynamic Drag 13
1.6 Aerodynamic Lift 14
1.7 Some Effects of Viscosity at High Reynolds Numbers 17
1.8 Stability and Control 21

Chapter 2. Rainfall and the Descent of Water Droplets **24**
2.1 The Physics of Rainfall 25
2.2 The Aerodynamics of the Sphere 27
2.3 Terminal Velocity 29
2.4 The Maximum Size of a Raindrop 31

Chapter 3. The Descent of Ice Crystals, Hailstones and Snow **34**
3.1 The Aerodynamics of Falling Ice Crystals 34
3.2 Terminal Velocity of Falling Ice Crystals 40
3.3 Optical Phenomena in the Atmosphere 44
3.4 Hailstones 47
3.5 Snow 52

Chapter 4. The Airborne Dispersal of Fruits and Seeds **53**
4.1 Dust Seeds, Spores, and Pollen 54
4.2 Plumed Seeds and Fruits 56
4.3 Plain Winged Seeds 57
4.4 Winged Fruits and Seeds which Autorotate 59
4.5 Tumbleweeds 67
4.6 An Overall View 67

Chapter 5. Bird Flight: I-Gliding and Soaring Flight **68**
5.1 Anatomy and Plumage 69
5.2 Basic Aerodynamics 71
5.3 Geometrical Scaling Laws 78
5.4 Gliding Flight 79
5.5 Diving Flight 83
5.6 Slope Soaring 84
5.7 Thermal Soaring 85
5.8 Dynamic Soaring 88

Chapter 6. Bird Flight: II-Flapping Flight **92**
6.1 The Aerodynamics of the Wing in Fast Forward Flight 92
6.2 The Aerodynamics of the Wing in Slow Forward Flight 95
6.3 Horizontal Flight 98
6.4 Characteristic Flight Speed 99
6.5 Flight Capability and the Influence of Weight 100
6.6 High-speed Flight 102
6.7 Hovering Flight 103
6.8 Take-off and Landing 106
6.9 Intermittent Flight Styles 107
6.10 Migratory Flight 112
6.11 Bird Flight at Altitude 115
6.12 Formation Flying 116
6.13 Vortex Theory of Bird Flight 120

Chapter 7. Insect Flight **123**
7.1 Anatomy 123
7.2 The Aerodynamics of Insect Flight 128
7.3 The Flight Speed of Insects 132
7.4 The Forward Flight of Large Insects 134
7.5 The Flight of Tiny Insects 135
7.6 Swarming Flight 136
7.7 The Clap–Fling Mechanism 138
7.8 Other Methods of Lift Generation 140

Chapter 8. The Flight of Bats, Pterosaurs and Other Vertebrates **143**
8.1 Bats 143
8.2 The Pterosaurs 149
8.3 Gliding Animals 153
8.4 Flying Fishes 154

References **156**
List of Symbols **159**
Glossary of Terms **161**
Further Reading **167**
Index **169**

Preface

Aerodynamics is primarily regarded as a discipline of engineering and the applied sciences. Yet aerodynamic phenomena play a significant role in several fields of study within the natural sciences. Examples are: the descent through the atmosphere of precipitation particles, such as rain, hail and snow; the airborne dispersal of seeds and fruits; and the flight of birds, insects, and bats. Biophysical aerodynamics is concerned with the scientific study of these topics within the natural sciences. Because biophysical aerodynamics makes contact with and overlaps into scientific disciplines such as meteorology, botany, zoology, ornithology, and entomology, cutting across the usual boundaries of these distinct areas of study, it provides a powerful vehicle for the cross-fertilization of ideas which have a common theme in aerodynamics.

Diaries and other records of pioneers in the field of aeronautics—men such as Leonardo da Vinci, Sir George Cayley, and the Wright brothers—show that careful observations and interpretations of the flight of birds and the airborne dispersal of seeds formed the foundation from which man first gained an appreciation of how mechanical flight could be achieved. It was in this aeronautical context that the increased understanding of aerodynamics rapidly expanded during and beyond the first half of the twentieth century. As a result of these advances, many aspects of biophysical aerodynamics can now be explained in a manner that would not have been possible in the nineteenth century. The wheel of learning has turned full circle. Yet the aerodynamicist has no room for complacency. The emphasis given to particular lines of enquiry in the world of aviation led to the situation in which conventional aerodynamic theory was incapable of explaining some of the complex, unsteady aspects of animal flight, a factor which fuelled independent explorations in the understanding of aerodynamic principles. This background led to the discovery in 1973, by the distinguished zoologist the late Professor Torkel Weis-Fogh, of a mechanism of lift generation, now known as the clap–fling mechanism, which was hitherto unknown to aerodynamicists. Further, in recent years, the development of the vortex theory of flight has been strongly motivated by the need to find an improved physical description of animal flight. It will be interesting to see whether, in the fulness of time, new engineering applications will be found for these advances in scientific thought.

In the main, this book adopts a descriptive, physical approach to the topics under consideration. Nevertheless, aerodynamics is a numerate endeavour, and so simple mathematical analyses are presented where appropriate. A glossary of aerodynamic terms is included. For those wishing to pursue particular matters of interest beyond the level that the present treatment allows, suggestions for further reading are given at the end of the book. Throughout the volume references are quoted where they are thought to be useful, but no effort has been made to be exhaustive in this respect.

The layout of the book reflects a logical development from the simple, drag-dominated, passive, movement of spherical water droplets, through a variety of stages of increasing complexity, to the active flight of animals capable of controlling the geometry of their aerodynamic surfaces and able to generate thrust and lift, as well as suffering the unavoidable consequences of drag. The first chapter is intended to provide, for the non-specialist, a general introduction to the study of aerodynamics. Chapters 2 and 3 are devoted to aerodynamic aspects of precipitation. The speed of descent of raindrops, hailstones, and snow are considered. Airborne seed dispersal is the subject of Chapter 4, whilst Chapters 5 to 8 are concerned with animal flight. Chapters 5 and 6 consider non-flapping and flapping flight in birds, respectively; the aerodynamics of insect flight is covered in Chapter 7, and the final chapter deals, amongst other topics, with the flight of bats and pterosaurs.

For permission to use previously published material, I should like to express my gratitude to the authors and publishers concerned. Specific acknowledgements are incorporated elsewhere. Throughout the preparation of the book my wife, Barbara, has made a huge contribution, not only in preparing the typescript from my handwritten notes but in many other ways as well.

This book is aimed at a wide audience. I particularly hope that readers will delve beyond the confines of their own specializations. To those readers who may have misgivings about such a venture the words of Alfred, Lord Tennyson, in *Ulysses*, are a fitting sentiment with which to conclude the Preface and enter upon the main body of the book:

> Come, my friends, 'Tis not too late to seek a newer world.

Before you write about creatures which can fly make a book about the insensible things which descend in the air without the wind and another on those which descend with the wind.

Leonardo da Vinci (1452–1519)

There is never a discovery made in the theory of aerodynamics but we find it adopted already by Nature.

Sir D'Arcy Thompson (1860–1948)
in *Growth and Form*

Chapter 1

General Introduction to Aerodynamic Principles

We live at the bottom of a vast ocean of air which entirely surrounds the earth. On a still day it is easy to remain totally unaware of the presence of this invisible medium. In our everyday life it is only when we become conscious of our breathing, or when we think how a passing aeroplane remains aloft, or a strong wind blows up, that the presence of this environment of air is brought to our attention. Yet this seemingly tenuous medium, besides providing oxygen, essential to the support of life, has another separate and crucial role to play in the natural world about us. The flight of birds, insects, and bats depends upon the generation of aerodynamic lift; in the plant world numerous species exploit aerodynamic phenomena to ensure that one generation succeeds another; and in the skies above us the clouds are composed of tiny water droplets or ice crystals, their motion dependent upon the aerodynamic forces to which they are subjected, and eventually to be precipitated in the form of rain, hail or snow.

These are some of the topics which we shall be discussing in this book. However, before considering such matters in detail, an introduction to the basic principles of aerodynamics is, perhaps, in order.

Our purpose in the opening chapter is to discuss briefly some of the essential elements of aerodynamics and, in particular, we have in mind the needs of those who are unfamiliar with the aeronautical world. Nevertheless, it would seem inappropriate in the present context to provide an extended introductory discourse on the subject, lest the whole balance of the book be distorted. For those who feel they would benefit from further background reading the following references are offered: *The Science of Flight* (Sutton, 1949); *Shape and Flow* (Shapiro, 1961); *Aerodynamics for Engineering Students* (Houghton and Carruthers, 1982); *Aerodynamics: The Science of Air in Motion* (Allen, 1982).

To further assist the reader a Glossary of Terms is to be found at the end of this book.

1.1 PHYSICAL PROPERTIES OF AIR AND THE ATMOSPHERE

All matter is composed of vast numbers of molecules surrounded by empty space. Air is no exception. Under typical sea level conditions there are about 3×10^{25} molecules in 1 m^3 of air. However, only the very smallest particles moving in the atmosphere behave in a way which reflects the molecular nature of the air surrounding them. The tiniest pollen grains and spores, and the minutest water droplets and ice crystals, may be just small enough to move randomly through the atmosphere, exhibiting Brownian motion as a result of the buffeting they receive from the motion of the individual molecules of air. But more generally the bodies with which we shall be concerned are too large for this effect to happen.

A characteristic length at the molecular level is very small compared with the smallest dimensions characterizing the geometry of bodies which are the subjects of study in the aerodynamics of the natural environment. For the conditions which exist within the atmosphere the mean free path of air—that is the distance that on average a molecule travels before colliding with another molecule—is of the order of 10^{-7} m, and the average distance between the centres of neighbouring molecules is about 3×10^{-9} m. By way of contrast the smallest particles with which the present work is concerned have a characteristic length of about 10^{-6} m. So the study of natural aerodynamics may be pursued by ignoring the molecular nature of the air and, instead, viewing it as a continuum having physical properties corresponding to the statistical average conferred by the molecules. This approach allows us to express the properties of the air in terms of such familiar quantities as pressure, temperature, etc. Thus the study of biophysical aerodynamics, both by way of experimental observation and theoretical analysis, is concerned with the motion of air whose physical properties are assumed to be continuous functions of space and time. This idealization of the physical nature of matter is known as the continuum hypothesis and underlies all the work considered in the following pages.

Density

The basic definition of the density of a substance is the ratio of the mass, m, of a given amount of the substance to the volume, $\mathcal{V}$, it occupies. Since air is compressible, this definition requires further clarification, however, and the point at issue may be highlighted in the following way.

Treating the air as a continuum, and using the limiting processes of differential calculus, our first instinct might be to define the density ρ as

$$\rho = \lim_{\delta\mathcal{V}\to 0} \frac{\delta m}{\delta\mathcal{V}} = \frac{dm}{d\mathcal{V}}$$

However, our knowledge of the molecular nature of air should make us aware that this definition is unacceptable, because the number of molecules occupying a given volume continually changes, and for small volumes this effect is most important. So, the elemental volume $\delta\mathcal{V}$ cannot be allowed to become indefinitely small, because under such circumstances our definition breaks down. Equally, because in practice the density of the air changes throughout the atmosphere, an unrealistically large volume cannot be used as the basis of the definition. There is an answer to the dilemma, and it is found as follows. We find a limiting volume $\delta\mathcal{V}^*$ such that it is sufficiently large for variations due to molecular movement into and out of the volume to have no significant effect on the density, yet, from the point of view of continuum theory the volume is so small that it does not impair our notion that the density of the air can vary continuously with respect to position within the atmosphere.

Hence, formally, the density, ρ, is defined by the relation

$$\rho = \lim_{\delta\mathcal{V}\to\delta\mathcal{V}^*} \frac{\delta m}{\delta\mathcal{V}} \tag{1.1}$$

where $\delta\mathcal{V}^*$ is the limiting elemental volume. As an example, consider air at typical sea level conditions. If we select a value for $\delta\mathcal{V}^*$ of 10^{-18} m^3, then this tiny volume contains about 3×10^7 molecules of air, even though the sides of the volume are only about 10^{-6} m long.

The equation of state

For the conditions of temperature and pressure which prevail in the atmosphere, air satisfies both Boyles' and Charles' Laws and, to all intents and purposes, behaves as a perfect gas.

The relationship between the density, pressure, and temperature of a gas is expressed through the equation of state and, for a perfect gas, this has the simple form

$$p = \rho R T \tag{1.2}$$

where p is the absolute pressure, T is the absolute temperature and R is the gas constant for air*, equal to 287 J kg^{-1} K^{-1}.

The standard atmosphere

From our everyday experiences we know that the ambient temperature changes through the day and with the seasons. Similarly the barometric pressure rises and falls continuously. By collecting records over long periods of time meteorologists have been able to determine the average conditions of temperature and pressure, at sea level and at altitude, in different parts of the world, and on a monthly or seasonal basis. This sort of information has many uses but so far as the aerodynamicist is concerned it is not ideal. In order to compare the performance of one aircraft with another an aerodynamicist requires a common base on which to make the comparison. This is the reasoning behind the concept of the standard

* Following common engineering practice R is used here to denote the specific gas constant for a particular gas. Care should be exercised here, however, for in some scientific disciplines this symbol is used to denote the universal gas constant.

atmosphere. The standard atmosphere is a mathematical model, which incorporates conditions of temperature and pressure broadly consistent with annual averages measured in a temperate climate, both at sea level and at altitude.

A standard atmosphere is constructed on the assumption that the air is still, and the atmospheric properties are defined by the temperature–height profile, which consists of a series of connected segments in each of which temperature varies linearly with height.

The mathematical basis of the standard atmosphere will be described briefly. Consider a small cylindrical element of air, of cross-sectional area A_c and height δz, at an altitude z, as in Fig. 1.1. The weight of air in the element must just balance the force resulting from the change in atmospheric pressure over the distance δz, and the equilibrium of the element is expressed by the relation

$$A_c((p + \delta p) - p) = -\rho g A_c \delta z$$

or, on simplification

$$\delta p = -\rho g \delta z$$

Taking the limit we derive the differential relation

$$\frac{\mathrm{d}p}{\mathrm{d}z} = -\rho g \tag{1.3}$$

where g is the acceleration due to gravity and the density ρ satisfies the state equation (1.2). Combination of equations (1.2) and (1.3) yields

$$\frac{\mathrm{d}p}{p} = -\frac{g}{RT}\,\mathrm{d}z \tag{1.4}$$

The integration of equation (1.4) depends upon the functional relationship between T and z.

The atmosphere is assumed to consist of a number of layers, within each of which the temperature varies linearly with height such that

$$T = T_n + \lambda_n(z - z_n) \tag{1.5}$$

In equation (1.5), T_n is the temperature at the height z_n at the base of layer n, and λ_n is the temperature gradient within that layer.

Substitution of equation (1.5) in equation (1.4) and subsequent integration yields one of two solutions. For a layer where λ_n is finite there results

$$\frac{p}{p_n} = \left(\frac{T}{T_n}\right)^{-g/R\lambda_n} \tag{1.6a}$$

whereas for a layer where λ_n is zero there results

$$\frac{p}{p_n} = \exp\left(-g\frac{(z - z_n)}{RT_n}\right) \tag{1.6b}$$

where, in both cases, p_n is the pressure at height z_n.

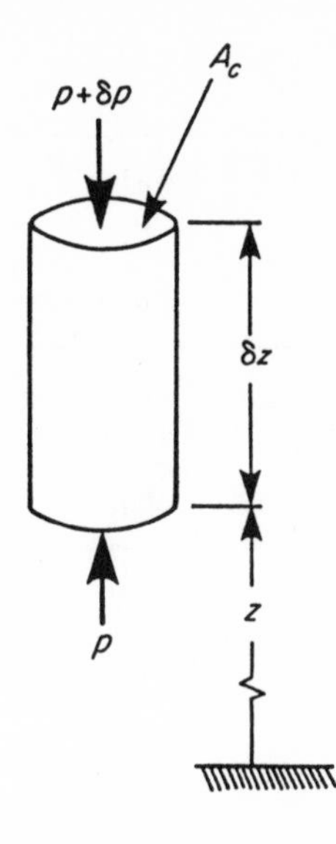

Figure 1.1. Forces acting on a small cylindrical element of air in the atmosphere.

Of particular importance is the International Standard Atmosphere (ISA). In the altitude range $0 < z < 11{,}000$ m the temperature–height relation for the ISA is

$$T = 288.15 - \lambda z \tag{1.7a}$$

where $\lambda = 0.0065$ K m^{-1}.
In the altitude range 11,000 m $< z <$ 20,000 m the temperature in the ISA is constant at a value

$$T = 216.65 \text{ K} \tag{1.7b}$$

The relationship between pressure and height is found by substituting equations (1.7a) and (1.7b) in equations (1.6a) and (1.6b) respectively. The density is then determined from equation (1.2).

The properties of the ISA up to an altitude of 15,000 m are summarized in Table 1.1.

Viscosity

When relative motion takes place between adjacent layers of a fluid (including both liquids and gases) internal shear stresses are set up which oppose that motion. This property of resistance to shearing motion due to the shear stresses is termed viscosity.

Dynamic viscosity

In order to quantify the effects of viscosity it is convenient to consider the flow which arises when a flat plate is moved slowly with a velocity, U, parallel to itself past a stationary wall. This flow is known as Couette flow and is illustrated in Fig. 1.2. There is no relative motion between the fluid and the solid boundary at their

Table 1.1 The international standard atmosphere (ISA).

Altitude z (m)	Temperature T (K)	Temperature t (°C)	Pressure p ($N\ m^{-2}$)	Density ρ ($kg\ m^{-3}$)
0	288.15	15	101,325	1.2250
1000	281.65	8.5	89,875	1.1116
2000	275.15	2	79,495	1.0065
3000	268.65	− 4.5	70,109	0.90912
4000	262.15	−11	61,640	0.81913
5000	255.65	−17.5	54,020	0.73612
6000	249.15	−24	47,181	0.65970
7000	242.65	−30.5	41,061	0.58950
8000	236.15	−37	35,600	0.52517
9000	229.65	−43.5	30,742	0.46635
10,000	223.15	−50	26,436	0.41271
11,000	216.65	−56.5	22,632	0.36392
12,000	216.65	−56.5	19,330	0.31083
13,000	216.65	−56.5	16,510	0.26548
14,000	216.65	−56.5	14,102	0.22675
15,000	216.65	−56.5	12,045	0.19367

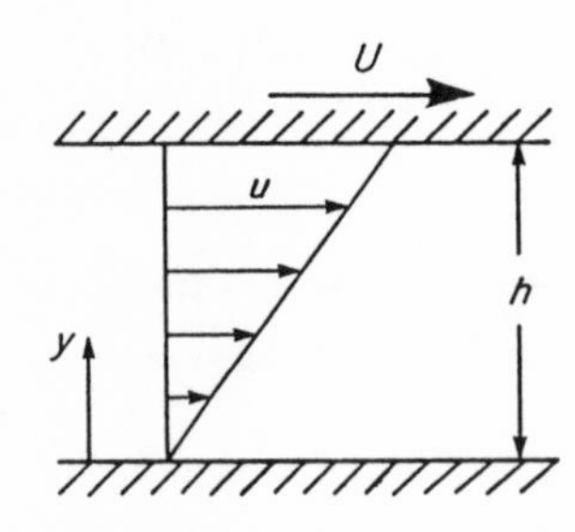

Figure 1.2. Couette flow, which provides a simple basis for the definition of dynamic viscosity.

common surface, so the fluid has zero velocity at the stationary wall and a velocity equal to U at the moving plate. The fluid exhibits a linear variation of velocity u with distance y from the stationary wall so that

$$u = \frac{y}{h} U \tag{1.8}$$

where h is the distance between the two surfaces.

In order to maintain the motion it is necessary to apply a tangential force to the moving plate. Equally a tangential force of the same magnitude is required to maintain the stationary wall at rest. The applied force is directly proportional to the area of the plate so it is convenient to consider the force per unit area of plate. This applied tangential stress is in balance with the viscous stresses in the fluid and all the layers of the fluid are subject to the same tangential shearing stress, τ.

By considering each element of the fluid it follows that τ is proportional to $\mathrm{d}u/\mathrm{d}y$, the velocity gradient normal to the plane of shear, and known as the rate of shear. Thus the shear stress can be expressed in terms of the rate of shear by the relation

$$\tau = \mu \frac{\mathrm{d}u}{\mathrm{d}y} \tag{1.9}$$

where the constant of proportionality, μ, is known as the dynamic viscosity.

The variation of the dynamic viscosity of air with temperature is shown in Fig. 1.3. For the range of pressures encountered in the atmosphere, the dynamic viscosity is independent of pressure.

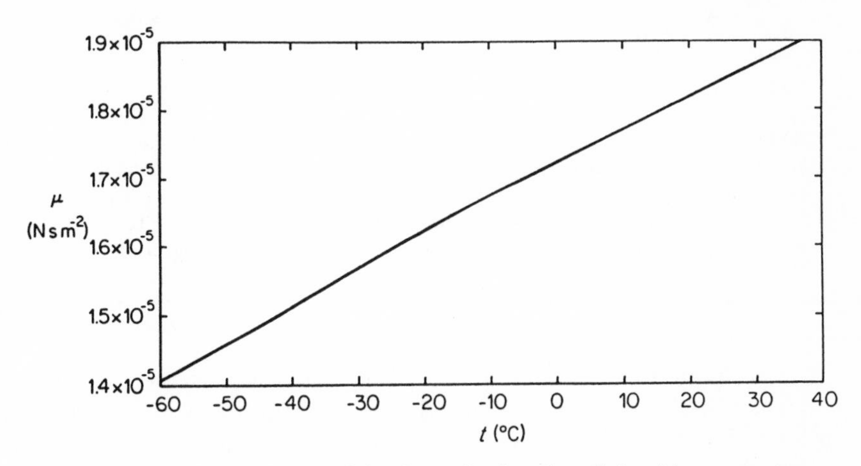

Figure 1.3. The variation of the dynamic viscosity of air with temperature.

Kinematic viscosity

When inertial and viscous forces are considered simultaneously, the ratio of the dynamic viscosity, μ, to the fluid density, ρ, often appears in such studies. This ratio is known as the kinematic viscosity, ν, and is defined by the relation

$$\nu = \frac{\mu}{\rho} \tag{1.10}$$

1.2 RELATIVE MOTION

It is useful, at an early stage, to introduce the idea of relative motion. As an example, consider a bird flying horizontally through still air with a speed of, say, $10\ \mathrm{km\ h^{-1}}$. If we stand still to observe this motion, we see the bird to be moving and the air to be still. But suppose, instead, that we are in a car or on a bicycle and can move along at the same speed and in the same direction as the bird. Then, provided we concentrate our attention on the bird (and ignore all extraneous factors in our field of view), the bird appears to us to remain stationary and we sense the air is moving towards the bird at 10 km h^{-1}. The important principle embodied in this example is that of relative motion. The velocity of the bird relative to the air is equal and opposite to the velocity of the air relative to the bird. It is often convenient when discussing the relative motion between a moving object and the air to regard the object as stationary and the airstream to be moving. As we have seen, the two situations are entirely equivalent, and in our subsequent discussions we shall feel quite free to choose whichever point of view we prefer (Fig. 1.4).

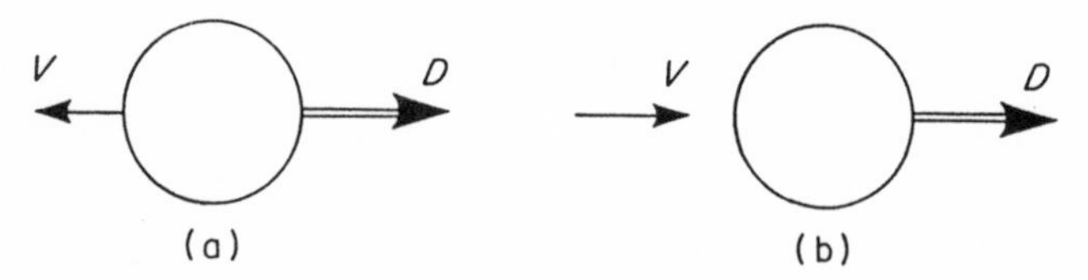

Figure 1.4. Relative velocities: (a) air at rest, ball moving right to left with velocity V; (b) ball at rest, air moving from left to right with velocity V. The two situations are, from an aerodynamic viewpoint, equivalent. In both cases the ball experiences a drag force acting towards the right, of magnitude D. The pattern of the airflow about the ball is the same in both cases.

The wind-tunnel, an important item of equipment for the experimental aerodynamicist, exploits this principle. In this device the relative motion between the air and a moving body is studied by holding the object still and allowing a stream of air to move past it.

1.3 REYNOLDS NUMBER

We have already seen that when moving, fluids—liquids and gases—generate internal shear forces which tend to resist that motion. In certain substances—treacle and thick oils are examples—this property of viscosity is particularly obvious, but all fluids, including air and water, exhibit viscous effects to a greater or lesser extent. Compared with liquids such as oils, air has a low viscosity but, notwithstanding this fact, the viscous nature of air has a vital role to play in aircraft aerodynamics and in the aerodynamics of the natural world.

Besides viscous forces there are other important forces in aerodynamics with which we are already familiar through our everyday experiences. We know that the harder the wind blows, the more firmly must we hold on to our umbrella to prevent it being snatched from our hand; the greater the strength of the wind, the greater the risk of our garden fence being blown down. These examples are manifestations of the inertial forces associated with the movement of air. The magnitude of the inertial force varies as the square of the windspeed and is directly proportional to the density of the air.

The pattern of the airflow about any object, and in consequence the resistance to motion that it experiences, depends very much upon the relative magnitudes of the inertial and viscous forces which arise within the flow. This important fact was first recognized by the famous nineteenth-century English engineer, Osborne Reynolds, who introduced a dimensionless number, now known as the Reynolds number, to quantify these effects. The Reynolds number, usually denoted by the symbol Re, is a measure of the ratio of the inertial forces to the viscous forces within a flow. It is evaluated by multiplying the airspeed relative to the object, V, by a characteristic dimension of the object (for example, the diameter d of a sphere, or the length l of a flat plate) and dividing the result by the kinematic viscosity, ν, of the air. For air under typical sea level conditions $\nu = 0.15 \times 10^{-4}\ \mathrm{m^2\ s^{-1}}$. Broadly speaking it can be said that in low Reynolds number flows ($Re < 1$), viscous forces predominate; at high Reynolds numbers inertial forces are predominant (but viscous forces cannot be entirely neglected).

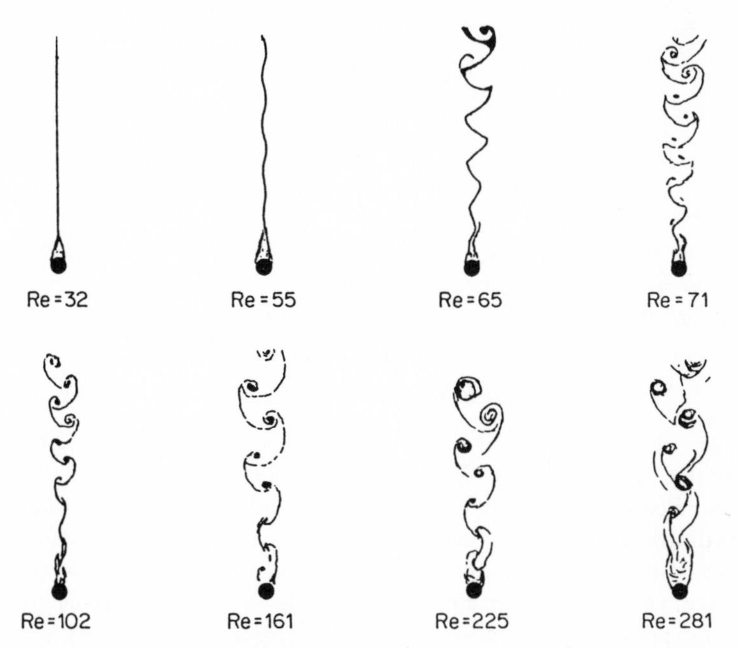

Figure 1.5. Flow patterns about a long circular cylinder, fixed normal to the airstream. The important influence of Reynolds number $Re = VD/\nu$ is evident. The airstream is flowing vertically upwards.

A simple visual impression of the importance of Reynolds number is given in Fig. 1.5, which shows the differing flow patterns which arise in the flow about a stationary circular cylinder for a range of Reynolds numbers. This diagram is relevant to the aerodynamics of falling ice crystals, a topic which is considered in Chapter 3.

The aerodynamics of the natural world encompasses a vast range of Reynolds numbers (Fig. 1.6). The motion through the air of tiny cloud droplets and insects

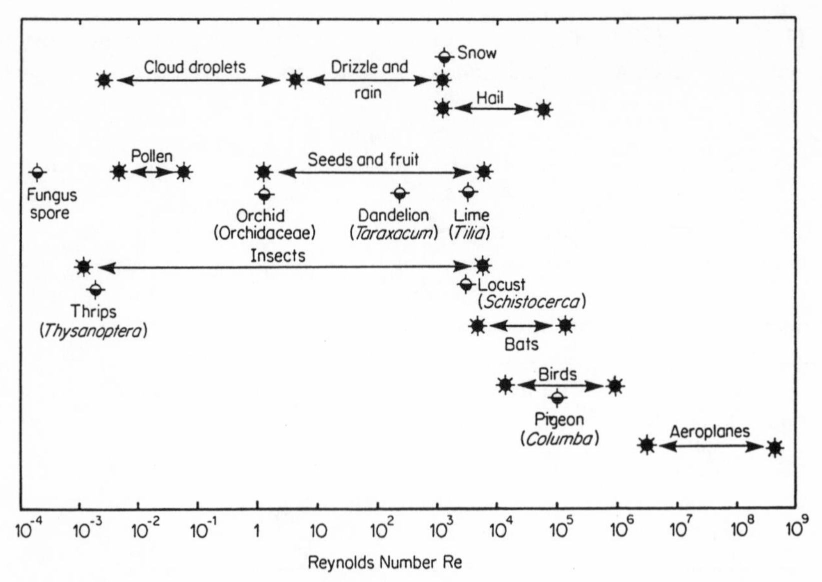

Figure 1.6. An indication of the range of Reynolds numbers with which the subject of biophysical aerodynamics is concerned. The range of Reynolds numbers at which aircraft operate is included for comparison.

is characterized by Reynolds numbers of the order 10^{-3}, whilst the largest birds fly at Reynolds numbers approaching 10^6. For the purpose of comparison, we note that aircraft operate at Reynolds numbers as high as 10^9.

1.4 THE FLOW FIELD ABOUT A MOVING BODY

When any body—whether it be a hailstone, a bird, or an aeroplane—moves through the atmosphere, certain changes are caused to take place in the air near the moving body. The distance away from the body to which perturbations of a substantial magnitude are communicated is directly related to the size of the body, and in directions perpendicular to the direction of movement these disturbances will only be significant up to a distance of a few body-lengths away. For even the largest bird (representative of the largest dimensions with which the subject of biophysical aerodynamics is involved) such length scales are tiny in comparison with the length scales of the atmosphere as a whole. Consequently, although the pressure and density of the atmosphere vary with altitude, when focusing attention on any aspect of biophysical aerodynamics no significant error is incurred if the motion is assumed to take place in a medium having a constant pressure and constant density corresponding to the altitude of the centre of mass of the moving object.

However, although the magnitudes of the disturbances caused to the air by a moving body diminish rapidly with distance, it is important not to overlook the fact that the integral effect of these small quantities can still be significant.

Consider, for example, a bird (or aeroplane) in horizontal flight. Whatever the altitude of flight, there is a reaction, at the ground, equal and opposite to the weight of the bird (or aeroplane), just as surely as if the bird (or aeroplane) was resting on the ground. This reaction is brought about by minute variations in atmospheric pressure spread over a large area at ground level; variations which are so small as to be unnoticeable without sensitive measuring equipment, yet having an integral effect which can add up to a very sizeable force!

Now that we have established the above principles let us move on and, for convenience, using the principle of relative motion, consider a stationary body immersed in a moving airstream. Far away from the body, in all directions, the air has a uniform velocity, V, say, as it is for all practical purposes undisturbed by the presence of the body in the airstream. Closer to the body the airstream must deviate from its original path in order to pass over the surface of the body, and this is reflected by the velocity of the air taking on values different from V. On the surface of the body the viscous nature of the air dictates that the velocity of the air relative to the body must be zero—the no-slip condition. The broad region which is influenced by the presence of the body in the moving airstream is known as the flow field. In the flow field not only does the velocity vary continuously in space, taking on values which, in general, differ from the freestream value, but, similarly, there is a corresponding distribution of pressure, which also differs from the undisturbed ambient condition. In contrast, the density of the air remains entirely unaltered by the relative motion of the air and body, at the low speeds appropriate to the study of biophysical aerodynamics. This fact was not appreciated by Leonardo da Vinci, who incorrectly attributed the lifting qualities of the wings of birds to a reduction in density (rarefaction) on the upper surface and an increase (compression) on the lower surface. Only at airspeeds approaching the speed of sound (330 m s^{-1} at sea level) do variations in air density become significant. Also, it should be noted that at low speeds the dynamic and kinematic viscosity of the air are unaffected by the relative motion of the air and body.

Thus, to summarize, despite the fact that air is compressible, and this quality is manifested by the density of the air varying with altitude throughout the atmosphere, it is possible to study the phenoma of biophysical aerodynamics as though the air were incompressible.

Under normal circumstances, when a body moves through still air, or an airstream moves past a stationary body, the pattern of movement of the air is invisible. To discover what is happening aerodynamicists carry out tests on objects, or accurate scale models, in wind-tunnels. Here they can use scientific instruments to measure velocities and pressures at any position within the flow field and on the surface of the object. Smoke can also be injected into the flow, and agents can be painted on to the surface of the body to gain an accurate visual impression of how the air moves about the body.

As is evident from the definition of dynamic viscosity (Fig. 1.2), viscous shear stresses are large where rates of shear are large; so by examining the distribution of velocity throughout the field of flow about a moving body it is possible to determine where shear stresses, and hence the manifestation of the air's viscosity, are important. At low Reynolds numbers, $Re \ll 1$, it is found that the rate of shear is of a reasonably uniform magnitude throughout the flow field, extending to a considerable distance in all directions away from the surface of the body. In this sense it is possible to state that viscous effects pervade the whole flow field, since they are no more nor less important in any one part of the flow rather than another. However, with increase in Reynolds number above 1, it is found that the velocity gradients become increasingly non-uniformly distributed, and the larger rates of shear become progressively confined to a narrow region close to the body and in a corresponding narrow region, known as the wake, downstream of the body.

At Reynolds numbers of about 10^3 and above, the large normal velocity gradients are restricted to a very thin layer of air, immediately adjacent to the surface of the body, known as the boundary layer, and in the wake. Outside of these regions the air behaves as though it were inviscid. For the purposes of mathematical analysis it therefore becomes possible to consider the flow field as divided into two separate regions. From a theoretical standpoint this is a matter of enormous importance because the inviscid flow region can be analysed using the powerful methods of potential flow theory.

Within the boundary layer, some of the terms appearing in the general equations of viscous flow theory can also be eliminated on order of magnitude grounds, so leading to the simplified equations which form the basis of boundary layer theory. However it is not our purpose to pursue this line of mathematical analysis further, and the interested reader should consult, for example, Schlichting (1979).

It is very handy to be able to convey a visual impression of the flow patterns about a moving body, and here the concept of the streamline is important. The streamline is a line which is conceived to lie within the flow, such that the local velocity vector is a tangent to the line. As a consequence of this definition no fluid flows across a streamline. In regions of high velocity the streamlines are close to each other; conversely, they are spaced well apart where the velocities are low.

We have already observed that the variation in velocity throughout the flow field is accompanied by a corresponding variation in pressure. At this stage it is worth noting that the pressure distribution impressed upon the surface of the body is of particular importance, and we shall refer to this distribution on a number of occasions in the following sections.

Referring to conditions at high Reynolds numbers, we note that the variation of pressure across a boundary layer is negligible and so the pressure distribution acting on the surface of a body is essentially the same as that of the 'inviscid' air just outside the boundary layer.

In the region outside the boundary layer and wake, the absence of significant viscous effects means that there is no agency by which mechanical energy can be degraded into thermal energy. Accordingly, the principle of conservation of energy may be applied to a small element of air in motion to show that there is a balance between the rate of increase of kinetic energy of the element and the rate at which pressure does work, since, for the aerodynamic conditions under consideration, changes in potential energy may be neglected as insignificant.

For the low-speed conditions of interest the air behaves as though it were incompressible and of constant density, ρ, and under these circumstances the energy balance can be represented by the simple equation

$$p + \frac{1}{2}\rho q^2 = \text{constant} = p_0 \tag{1.11}$$

where p_0 is known as the total or stagnation pressure, and q is the magnitude of the local velocity vector. For a Cartesian coordinate system, with the velocity components u, v, w in the directions of increasing x, y, z, respectively, then

$$q^2 = u^2 + v^2 + w^2.$$

Equation (1.11) is known as Bernoulli's equation, after the eminent Swiss mathematician Daniel Bernoulli (1700–1782). In particular the motion of the air along any streamline far upstream of the body must satisfy this equation, so that if p_a is the ambient pressure and V is the velocity of the undisturbed airstream relative to the body then

$$p_a + \frac{1}{2}\rho V^2 = p_0 = p + \frac{1}{2}\rho q^2.$$

1.5 AERODYNAMIC DRAG

Any object which moves through the air experiences an aerodynamic force which can be resolved into components in the direction of, and perpendicular to, the relative velocity vector. These components are known, respectively, as aerodynamic drag and aerodynamic lift or, simply, drag and lift.

For any body of arbitrary shape and having arbitrary orientation with respect to the airstream the drag consists of two components. The first is the normal pressure drag, which is the resultant force arising from the resolved components of the pressures acting normal to the surface of the body. The second is the skin-friction drag, which is the net force resulting from the resolved components of all the tangential forces, due to the viscous shear stresses, on the surface of the body.

These are general definitions, but to press the discussion forward it is convenient at this stage to concentrate upon the flow about bodies of arbitrary shape so orientated with respect to the flow direction that the lift is zero. For such conditions the drag force is, in general, made up of two components—the form drag and the skin-friction drag—whose sum is called the profile drag.

We may easily distinguish between these two components by considering the flow about a flat plate of negligible thickness (Fig. 1.7). When the plate is aligned with the airstream the profile drag is entirely due to skin-friction drag; if the direction of the plate is now turned through 90°, so that it is perpendicular to the

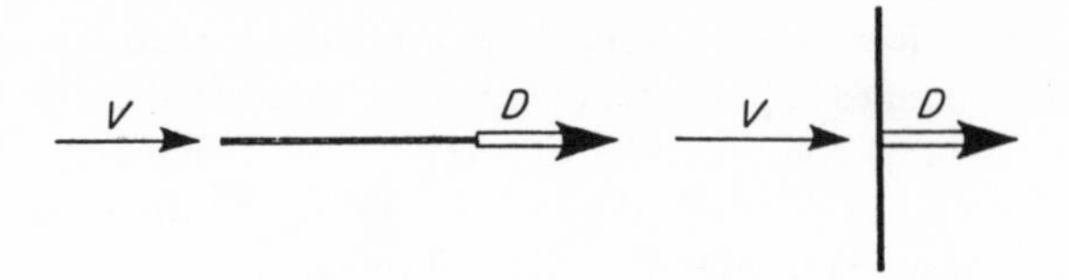

Figure 1.7. Profile drag of a flat plate: (a) skin-friction drag; (b) form drag.

incident airstream, then the profile drag is entirely due to form drag. For a body of more general shape both of these components of drag are present simultaneously. The form drag is, for a symmetrical body symmetrically disposed with respect to the airstream, identical to the normal pressure drag. For the flow about a streamlined shape the main contribution to the profile drag is skin-friction drag, whereas for an unstreamlined shape (known in aerodynamics as a bluff body) such as a sphere, form drag is the main component.

As an approximate guide we can state that for $Re < 1$ the drag is directly proportional to V, whereas at high Reynolds numbers (say $Re > 10^3$) the drag is approximately proportional to V^2. In the intermediate range the power of V in the drag law gradually changes from 1 to 2. It is emphasized that, at this stage, these statements are only intended as rough rules-of-thumb. In Chapters 2 to 8 we shall pay much greater attention to the variation of drag with Reynolds number for different body shapes, and will find that on occasions these rough-and-ready rules break down.

1.6 AERODYNAMIC LIFT

So far we have largely concentrated our attention on the flow about symmetrically shaped bodies, and in particular the flat plate and sphere. Let us now look in more detail at the flow about the flat plate. We have considered the drag force generated when the plate is aligned with, or is perpendicular to, the airstream. In both of these cases the flow pattern is symmetrical.* If, now, a small angle is established between the plate and the direction of the airstream, the plate is said to be at incidence relative to the airstream. The flow is no longer symmetrical and, in addition to a resistance force which acts in the direction of motion of the airstream, the plate experiences aerodynamic lift, in the direction perpendicular to the drag force (Fig. 1.8).

A flat plate of small thickness can be a reasonably efficient lifting surface; even more efficient are the wings of birds and aircraft, which may have quite a complicated planform geometry and cross-sectional shape. To generate high ratios of lift to drag a special cross-sectional shape, known as an aerofoil section, is required. Low-speed aerofoils have rounded leading edges and angled trailing

* There are certain ranges of Reynolds number for which the flow pattern is inherently unsteady and unsymmetrical, but for the purposes of the present discussion we can overlook such conditions.

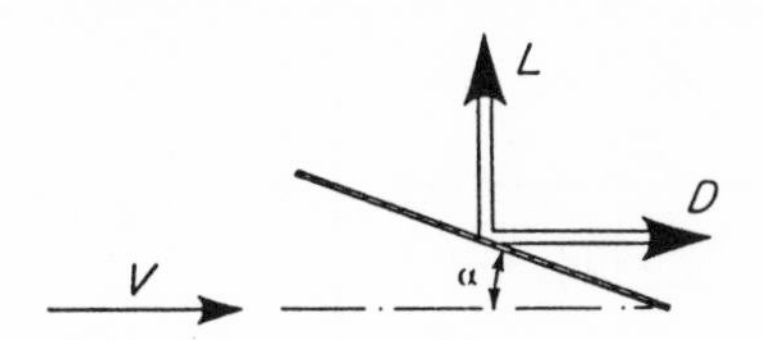

Figure 1.8. Flat plate at incidence. The plate experiences both lift, L, and drag, D.

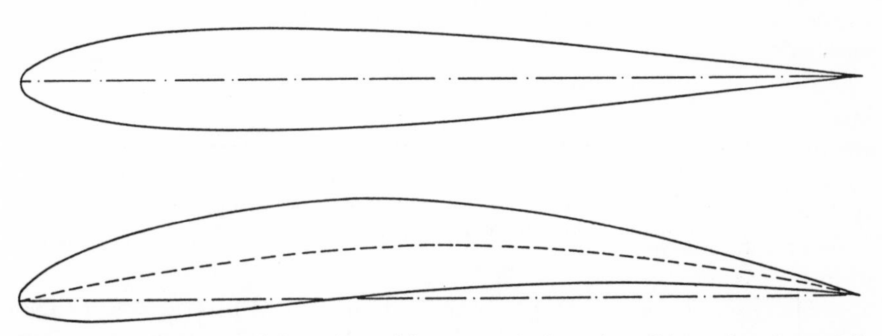

Figure 1.9. Basic aerofoil sections: (a) symmetrical section; (b) cambered aerofoil (— · — · —, chord line).

edges, and may be symmetrical or unsymmetrical (cambered) in section (Fig. 1.9). The line joining the centres of curvature of the leading and trailing edges is known as the chord line, and the angle of incidence is the angle between the chord line and the direction of the airstream. The distance between the leading and trailing edges, measured along the chord line, is defined as the chord, denoted by the symbol c.

Although, in principle, lift forces can be generated at any Reynolds number, the efficient production of aerodynamic lift, with attendant high lift/drag ratios, is a high Reynolds number phenomenon, and our discussion of aerodynamic lift will be restricted to such high Reynolds numbers.

All wings tend to display certain basic aerodynamic characteristics at the high Reynolds numbers at which they normally operate. Consider a wing with a symmetrical aerofoil section. When placed in an airstream with a constant airspeed V, if the angle of incidence, α, of the wing is systematically varied, the lift increases linearly with incidence until at a certain angle of incidence, round about 10–15° depending on the wing geometry, the lift reaches a maximum value. With further increase in incidence the lift diminishes. The drag of the wing increases with incidence, at first slowly, but then more rapidly (Fig. 1.10).

The increase in drag as the angle of incidence of the (symmetrical) wing is increased from zero is known as lift-dependent drag, and is mainly due to a new component of drag, known as the vortex or trailing-vortex drag (formerly called induced drag). This trailing-vortex drag is a direct consequence of the lift being generated by the wing and manifests itself on the wing as an additional contribution to normal-pressure drag. Although the main component of the

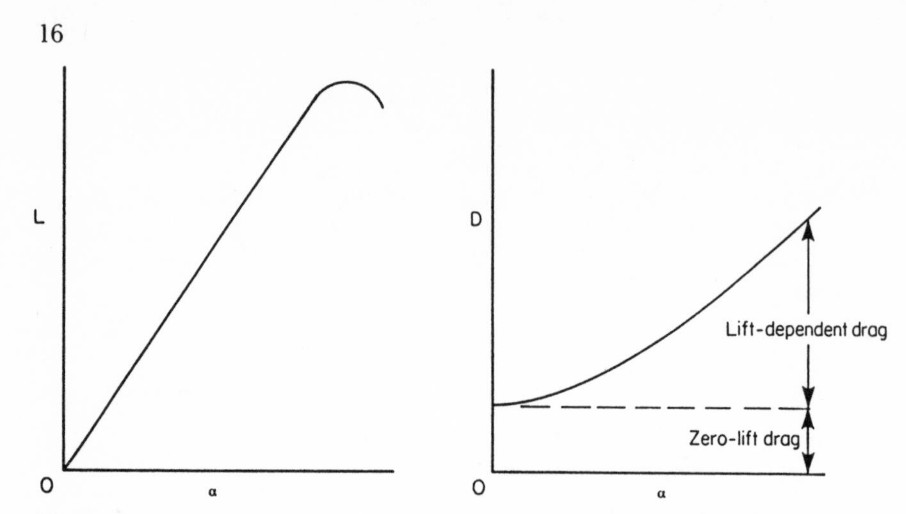

Figure 1.10. The variation of lift and drag of a wing with angle of incidence, at constant airspeed. Symmetrical aerofoil section.

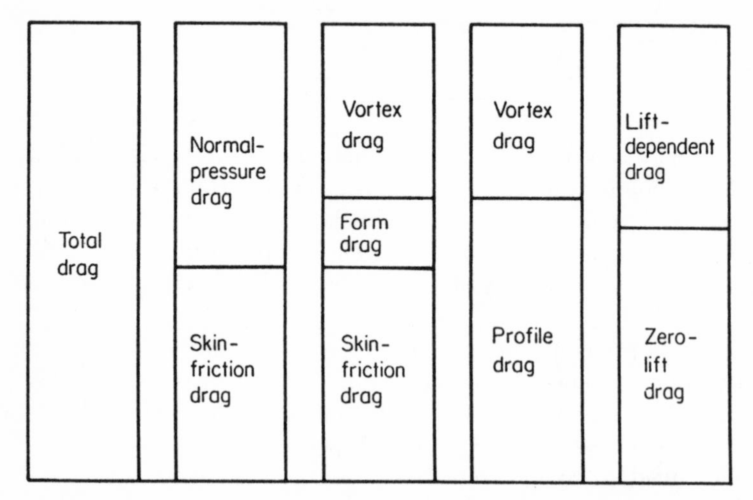

Figure 1.11. The components of drag of a lifting wing. It is shown that the total drag can be subdivided in several different ways.

lift-dependent drag is the trailing-vortex drag, there is also a significant contribution from the profile drag, which itself increases with incidence (Fig. 1.11). For the range of Reynolds numbers appropriate to the flight of aircraft and birds, (typically $Re > 10^4$) both the lift and drag forces increase in direct proportion to the wing area, S, and air density, ρ, and vary as the square of the airspeed, V. It is common practice to introduce the lift and drag coefficients, C_L and C_D respectively. They are defined by the equations

$$L = \frac{1}{2} \rho V^2 S C_L \qquad (1.12)$$

$$D = \frac{1}{2} \rho V^2 S C_D \qquad (1.13)$$

where L and D are the lift and drag, respectively.

The aerodynamic lift on a wing results from the acceleration of the air, just outside the boundary layer, over the surface of the wing. High velocities are achieved on the upper surface and these are associated with the creation of regions of low pressure, a consequence of Bernoulli's equation, equation (1.11). It is the low-pressure region on the upper surface of the wing which is primarily responsible for the generation of lift; depending upon the geometry of the wing sections and the angle of incidence of the wing the pressure distribution on the lower surface of the wing may provide modest assistance with the process of lift generation, or it may contribute a net downward force, tending to reduce the effectiveness of the upper surface (Fig. 1.12).

At the wing tips there is a constant tendency for the air to flow from below the wing into the low-pressure region above it and, as the wing moves through the air, a pair of contra-rotating vortices is left trailing behind the wing. It is the presence of trailing vortices which gives rise to the trailing-vortex drag, since the wing constantly has to do work to create the rotational kinetic energy of the vortex.

The interpretation of the aerodynamic lift in terms of the resolved components of the pressure distribution impressed upon the wing by the airflow is the most readily understood explanation of the physics of lift generation. However, aerodynamicists sometimes find it convenient to describe the flow in different, but equivalent, terms, and it is perhaps advantageous to introduce these alternative descriptions at this stage.

One of Newton's laws of motion states that to every action there is an equal and opposite reaction. Consequently the aerodynamic lift on the wing may be regarded as the reaction which is equal and opposite to the rate of change of downward momentum imposed upon the air by the wing.

Finally, there is yet another way of considering the process of lift generation. The acceleration of the flow over the upper surface of the wing, and the corresponding modification to the velocity of the air passing across the lower surface, could be obtained, in principle, by replacing the wing by an equivalent vortex, known as the bound vortex, of the appropriate strength. Aerodynamicists measure the properties of vortices using such terms as vorticity and circulation. Suffice it to say at this stage that the lift generated by this equivalent bound vortex is directly proportional to the circulation of the vortex. Hence to double the lift on a wing the circulation of the bound vortex must be doubled. The system comprising the two trailing vortices plus the bound vortex is often referred to as a horseshoe vortex system (Fig. 1.13).

1.7 SOME EFFECTS OF VISCOSITY AT HIGH REYNOLDS NUMBERS

When the detailed nature of the flow about a streamlined body—such as an aerofoil operating at moderate angles of incidence—is examined we see that at high

Reynolds numbers there is a thin region—the boundary layer—very close to the surface in which the velocity changes very rapidly with distance away from the surface of the body (Fig. 1.14). The thickness of the boundary layer increases with distance downstream from the leading edge of the body.

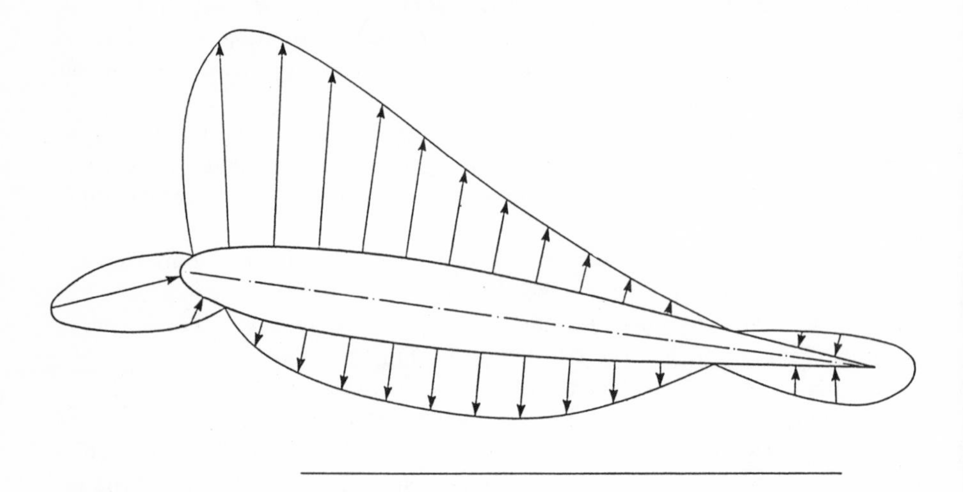

Figure 1.12. The pressure distribution acting on the surface of a wing at incidence. The precise form of the distribution depends on the wing geometry and angle of incidence.

The precise nature of the flow within the boundary layer depends on several factors, including surface roughness and the Reynolds number, Vc/ν, at which the aerofoil is operating. At Reynolds numbers, Vc/ν, of the order 10^5 or less, from the leading edge to the trailing edge and on the upper and lower surfaces, the adjacent layers of air within the boundary layer move without appreciable mixing, and the flow is said to be laminar. But at a higher Reynolds number, $Vc/\nu = 10^6$, say, a rather different situation emerges. Initially the boundary layers on both the upper and lower surfaces develop away from the leading edge in the form of laminar boundary layers. However, part way along the aerofoil surface the flow within the boundary layer starts to display signs of instability and further downstream, if examined in detail, the flow in the boundary layer exhibits a thoroughly disorganized appearance, fluctuating rapidly in space and time, with considerable inter-mixing of the flow. The flow is said to have become turbulent, and the process of change from laminar to turbulent motion is known as transition. At very high Reynolds numbers, transition takes place close to the leading edge and almost the entire boundary layer is turbulent.

All of the features that have just been described are clearly visible in the smoke rising above a lighted cigarette at rest in still air. Immediately above the tip of the cigarette the smoke rises regularly and in a vertical path; here the flow is laminar.

Higher up the smoke pattern starts to oscillate and the regular motion breaks down as transition occurs, and when the motion is completely irregular it has become fully turbulent.

On a thin aerofoil at modest angles of incidence the boundary layer remains attached to the surfaces of the aerofoil along its entire length to the trailing edge, where it leaves the aerofoil and forms a narrow wake.

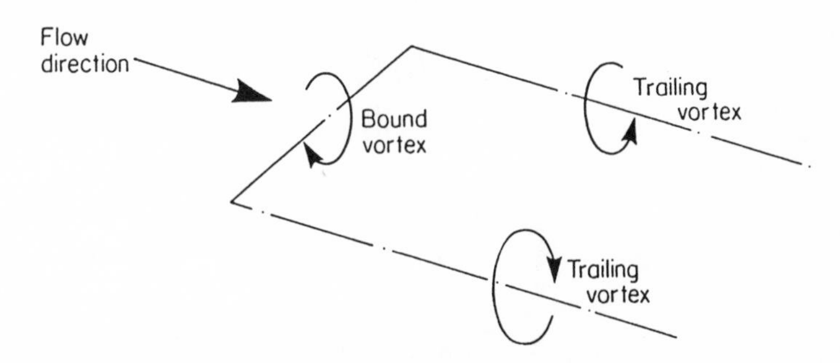

Figure 1.13. Representation of the flow field due to a wing by means of a pair of trailing vortices together with a bound vortex. This system is called a horseshoe vortex.

For a bluff body, or an aerofoil at an angle of incidence greater than that at which the maximum lift occurs, the situation is different. Close to the front of the body the boundary layer follows faithfully the contours of the surface, but there comes a stage in the development of the boundary layer on the upper surface of the aerofoil where it can no longer follow the profile on which it is growing, and the boundary layer separates from the surface. This process of separation gives rise to a broad wake with which is associated a high profile drag—in particular a high form drag.

The flow patterns about, and the forces acting upon, bodies immersed in an airstream can be quite complex. In order to improve our understanding of these matters it is helpful to consider a special concept, an ideal fluid which has no viscosity. By making mathematical calculations of the flow about bodies immersed in such an ideal fluid, and comparing the results with what happens in the real world, it is possible to gain a better understanding of the actual flow.

Some rather surprising results emerge from this type of analysis. Firstly, it is found that the drag of any body moving uniformly through an ideal fluid is zero. This result is often referred to as d'Alembert's paradox. Secondly, if an aerofoil section is set at an angle of incidence in a uniform stream it is found to generate zero lift in an ideal fluid. Two stagnation points, that is points within the flow where the air is brought to rest, occur on the surface of the aerofoil: one is near the leading edge, the second is on the upper surface of the aerofoil close to the trailing edge. At

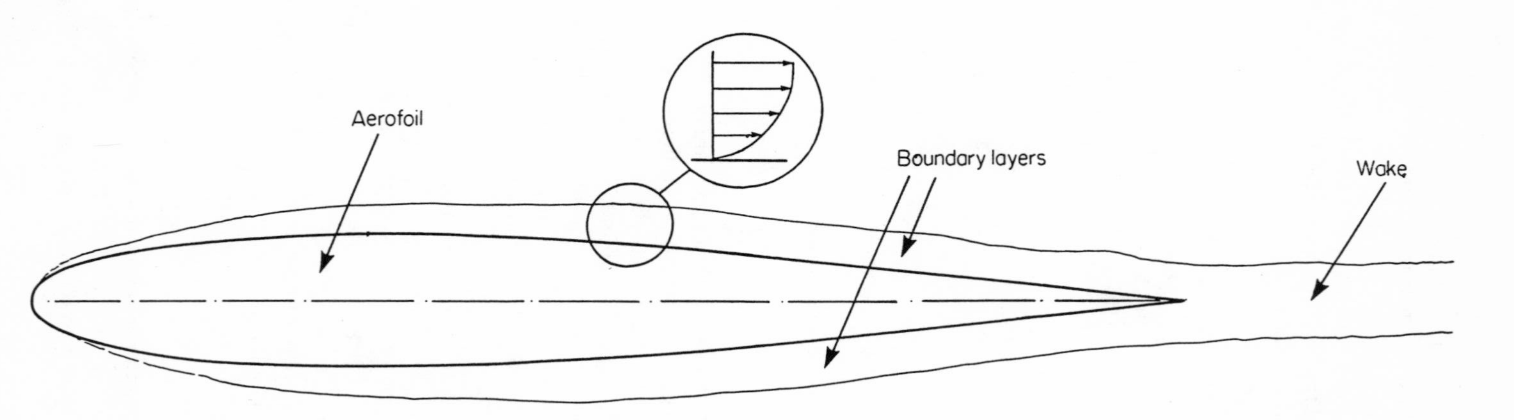

Figure 1.14. The boundary layers and wake (shown to exaggerated scale) in the flow field about an aerofoil at high Reynolds numbers. A diagrammatic representation of the velocity profile in the boundary layer is shown inset.

the trailing edge itself the mathematical analysis indicates that an infinite velocity would occur in an ideal fluid. To eliminate the inifinity at the trailing edge in the ideal fluid, and to produce a mathematical model which more closely fits the real world, it is necessary to introduce circulation round the wing; the appropriate circulation is determined by the requirement that it must be of such a strength as to move the rear stagnation point to the trailing edge of the wing. This important criterion, which determines the strength of the circulation about the wing, and in consequence the magnitude of the lift-force sustained by the wing, is referred to in aerodynamic circles as the Kutta–Joukowski condition.

In considering the flow of an ideal fluid we have been driven, almost artificially, to introduce the idea of circulation into the flow to account for the generation of lift and to produce a flow pattern more representative of a real flow.

It is natural to enquire about the source of this circulation in the real world, and we soon discover that its origins lie in the viscous nature of air. So, not only is viscosity responsible for the resistance to motion experienced by any body moving through the air, but also, almost paradoxically, the ability of this same body to generate aerodynamic lift is equally dependent upon the fact that the air is viscous.

1.8 STABILITY AND CONTROL

A body dependent upon aerodynamic effects to sustain its motion must develop the appropriate aerodynamic force, but this is not sufficient in itself. It is also necessary to consider the resultant system of forces acting on the body, in order to see that they satisfy certain stability criteria.

Consider a system subject to a number of forces which are in equilibrium. The system is said to possess the property of stability if, when the system is displaced by a small perturbation, it tends to return to its initial state on the removal of the perturbing effect. In direct contrast, an unstable system diverges away from the initial state when subjected to a small perturbation.

As examples of natural bodies which exhibit a high degree of inherent stability, we may quote the fruits of the dandelion (*Taraxacum*) and of the lime (*Tilia*). Whereas the dandelion uses drag for dispersal, aerodynamic lift is of fundamental importance to the lime. However, in both cases the correct orientation of the disseminule relative to the air through which it moves is achieved by the same principle, namely the low carriage of the seed. If for any reason the orientation of the fruit is disturbed from its preset path, for example by a sudden gust of wind, then the resulting lack of alignment of the aerodynamic and gravitational forces acting on the fruit gives rise to a turning moment which tends to restore the fruit to its original orientation. This is illustrated schematically for the dandelion fruit in Fig. 1.15. In this case the condition of stable equilibrium corresponds to the orientation for maximum aerodynamic drag, a desirable feature for a fruit relying on the mechanism of drag for dispersal.

A simple demonstration of the motion of a body lacking stability under the influence of aerodynamic and gravitational forces may be produced by attempting to launch a postcard into flight. All that results is a haphazard, indeterminate descent towards the ground, because there is no stable equilibrium state at which the falling postcard may settle.

In the above examples the geometry of the moving body is fixed and, if the body is capable of moving in a stable manner, it is an inherent property of its morphology. These simple stability considerations can be carried over and applied to certain aspects of animal flight. For example, during some (but not all) conditions of gliding and soaring flight in birds the flying surfaces are not moved because the equilibrium flight condition is inherently stable.

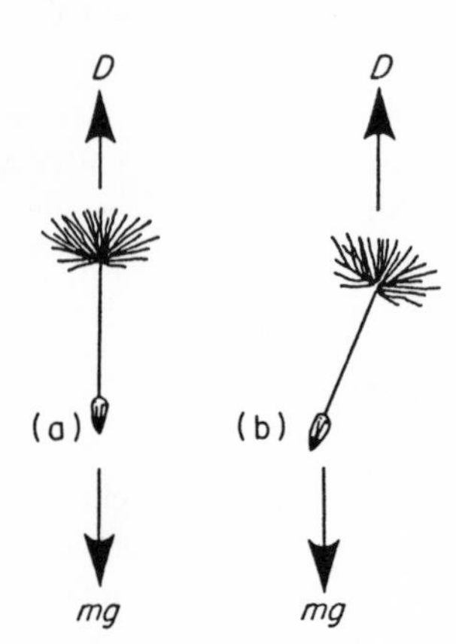

Figure 1.15 Dandelion fruit falling through still air: (a) condition of stable equilibrium—the forces of drag and weight are aligned; (b) the fruit has been disturbed from its equilibrium condition—the weight and drag are no longer aligned, but together they produce a turning moment tending to restore the fruit to its equilibrium condition.

As an introduction to slightly more advanced matters, let us consider the problem of a human balancing on one leg. The weight of the body is balanced by an equal and opposite reaction between the foot and the ground. By considering a small misalignment of these forces and applying the arguments used in discussing Fig. 1.15, we quickly recognize that this is an unstable configuration. Yet a person is quite capable of maintaining this unstable position for a considerable period of time. This is achieved in the following way. A human can sense that he is starting to fall over, and he can use this information to command his body to take corrective action, in this case by means of muscular activity to redistribute his weight, utilizing arm or body movements. Using this mechanism of feedback control the human is able to maintain a posture which is inherently unstable.

Feedback control systems of this kind are ubiquitous in nature. They can be used to good effect in those regimes of gliding and soaring flight which are inherently unstable. They play an even more important role in the highly complex and diverse forms of locomotion represented by active flight in birds, insects, and bats. The principles of stability and control, and their application to flight, form a substantial field of study, which stands in its own right. In this book we have insufficient space to do other than make occasional references to this topic.

Chapter 2

Rainfall and the Descent of Water Droplets

Whenever there are clouds in the sky it is a sure sign that water or ice particles are present in the atmosphere. The motion of these tiny particles, and of raindrops, hailstones, and snow, into which forms they ultimately grow, is vitally influenced by aerodynamic forces.

Initial formation of the particles comes about in the following way. Under certain meteorological conditions convection currents carry moist surface air to regions at high altitude where the air becomes supersaturated. If the temperature is above freezing, water condenses out from the moist air to form minute water droplets. When the convection currents are sufficiently strong, the air will be taken to yet higher altitudes where the temperature is well below freezing point. At these levels ice crystals are formed by the freezing of liquid droplets or by sublimation, that is the direct transformation of water vapour to the solid form, ice. A variety of fascinating crystal shapes are thereby created, the geometry depending upon the ambient temperature and the degree of supersaturation that exists. In temperate climates altitudes as high as 9000–15,000 metres (30,000–50,000 ft) might be required for the formation of ice crystals, whereas in the polar regions ice crystals can be present in the atmosphere close to ground level.

Once formed the ice crystals at high altitude and the water droplets at lower levels become subject to gravitational forces and, in the absence of updraughts, start to descend through the surrounding air under their own weight (the upward force due to buoyancy is negligible). In turn, the motion itself results in each particle experiencing an aerodynamic drag force. A stage is soon reached where the aerodynamic and gravitational forces acting upon the particle are in balance and its velocity becomes essentially constant.* Under these conditions the particle is said to be falling at its terminal velocity.

The size of water particles in the atmosphere embraces a range spanning from minute droplets, of which clouds are composed, up to the largest raindrops that can exist without breaking up. The diameter of these smallest droplets is about 10^{-6} m (4×10^{-5} in.), whereas raindrops do not occur at sizes above about 5.8×10^{-3} m (0.23 in.).

2.1 THE PHYSICS OF RAINFALL

The formation of water droplets rests in the first place upon the presence in the atmosphere of myriads of tiny particles; it is upon these particles that the process of condensation of supersaturated water vapour is initiated. A small proportion of the nuclei, of the order of one in a thousand, are hygroscopic and, because of their affinity for water, these droplets grow preferentially. The rate of growth due to condensation diminishes as the droplet increases in size and the largest droplets created purely by condensation are about 0.03 mm in diameter. Because of the non-uniform rates of growth of individual droplets a typical cloud contains droplets spanning a wide range of sizes. The larger droplets, being heavier, have higher terminal velocities than the smaller droplets, and this fact underlies the mechanism by which still larger droplets and raindrops are formed. For any falling water droplet a circular cylinder can be defined by the vertical path traversed by the drop and by its projected frontal area. If a large droplet is above a smaller one and the defining cylinders of the droplets overlap, it is likely that, sooner or later, the two droplets will collide during their descents. However, collision is by no means inevitable. Consider two water droplets of different sizes with terminal velocities V_D and V_d, respectively, descending vertically through the atmosphere. Whilst the droplets are some distance apart they move closer with relative velocity $(V_D - V_d)$. As they approach each other the smaller droplet becomes influenced by the flow field generated by the larger droplet, and there may be a tendency for the smaller droplet to be deflected from its vertical path. If there is a large disparity in sizes, a small droplet will tend to follow the streamlines around the larger droplet and a collision between the two may, as a consequence, be avoided. On the other hand if the two droplets are of a comparable size the inertia of the smaller droplet prevents any undue deviation from its initial path and a collision results in the creation of a single larger droplet. Thus a droplet increases in size as a result of successive collisions with other droplets until, after several collisions, the aggregation has given rise to a particle of sufficient size to be regarded as a raindrop.

If the defining cylinders of two droplets of similar size do not overlap but are, nevertheless, sufficiently close then a further mechanism exists by which the droplets can be brought into contact, provided the droplets are larger than about

* The velocity continues to vary slightly due to the change of density with altitude as the particle falls.

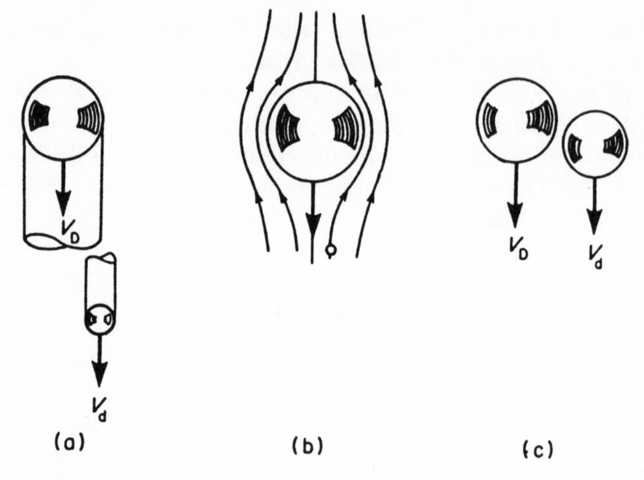

Figure 2.1. Some examples of the interaction between two falling rain droplets. (a) The defining cylinders overlap and, in general, a collision will occur from which a single larger droplet is formed. (b) A case where a collision may not take place. Although the defining cylinders overlap, the inertia of the tiny droplet is so small that it will tend to follow the streamlines around the larger droplet, so avoiding contact. (c) In this case, for droplet Reynolds numbers in excess of about 10, the defining cylinders of the two droplets do not overlap, although they are close to touching. If the two droplets are of a similar size, their relative velocity $(v_D - v_d)$ is small, and they mutually interact for a considerable period of time. A transverse force is created which may result in coalescence.

0.6 mm in diameter (below this size the Reynolds numbers of the droplets are too low for inertial forces to have any significant effect on the motion). The pressure on the surface of a single droplet in the equatorial region is lower than the ambient pressure, because the air is accelerated to its maximum velocity in this region. When two droplets of similar size move into close proximity there is a mutual interaction between the flow fields about the two droplets, which accentuates the low pressures in the equatorial regions where the two droplets move closest to each other; this is brought about by the further acceleration of the air in the small gap that exists. As a result of the asymmetrical pressure distribution so formed, each droplet becomes subject to a lateral force causing the paths of descent of the droplets to move towards each other. Provided the separation between the two droplets is not too great, and if their terminal velocities are similar, then the two droplets will finally coalesce to form a single droplet of increased size. If all the conditions required for contact to occur are not satisfied then the larger, faster-moving droplet will simply overtake the second droplet and they will go their separate ways. In the nature of things this mechanism of droplet formation due to coalescence is much rarer than that of formation as a result of collisions.

A second, distinct, process by which raindrops are created is as follows. When ice crystals are formed, as will be described subsequently, they may aggregate into snowflakes which, if they descend into warmer layers of the atmosphere, melt and precipitate in the form of raindrops.

The smallest water droplets in the atmosphere are almost perfectly spherical, as their shape is determined by surface tension. Even up to diameters of 1.5 mm, deviations from the spherical shape are only small. With increased size surface tension effects become progressively less important in comparison with aerodynamic forces, which increase approximately as the square of the terminal velocity. The pressure distribution associated with the aerodynamic forces will not be discussed in detail here. Suffice it to state that, whereas, relative to the freestream pressure, the raindrop suffers a reduction in pressure around the horizontal equator, the pressures at the front and, to a lesser extent, the rear of the raindrop are increased. This pressure distribution tends to squash the raindrop out, increasing its frontal area compared with that of a sphere of the same volume. This effect increases with the size of the raindrop. Hence the terminal velocity of a large raindrop is lower than it would otherwise be if the particle were purely spherical.

Since droplets up to 1.5 mm in diameter are essentially spherical, their motion can be computed using standard aerodynamic data for flow about spheres.

2.2 THE AERODYNAMICS OF THE SPHERE

Due to its complete rotational symmetry about all axes, a non-spinning sphere is one of the simplest shapes to deal with from an aerodynamic viewpoint, and the drag characteristics of spheres with smooth surfaces have been widely investigated and are well understood (Mason, 1978).

The drag force, D, can be written in the form

$$D = \frac{1}{2} \rho V^2 \times \frac{\pi d^2}{4} \times C_D \tag{2.1}$$

where C_D is the drag coefficient and the projected frontal area $\pi d^2/4$ has been used as the reference area.

The drag coefficient varies with Reynolds number, Re, in the manner shown in Fig. 2.2. The Reynolds number is defined by the relation

$$Re = \frac{Vd}{\nu} \tag{2.2}$$

where ν is the kinematic viscosity.

In spite of the simple shape of the sphere it is evident that the relationship between drag coefficient and Reynolds numbers is by no means straightforward. This fact is associated with the changing pattern of flow about the sphere. As has been pointed out previously, the Reynolds number is a measure of the relative importance of viscous and inertial forces within the flow about the sphere.

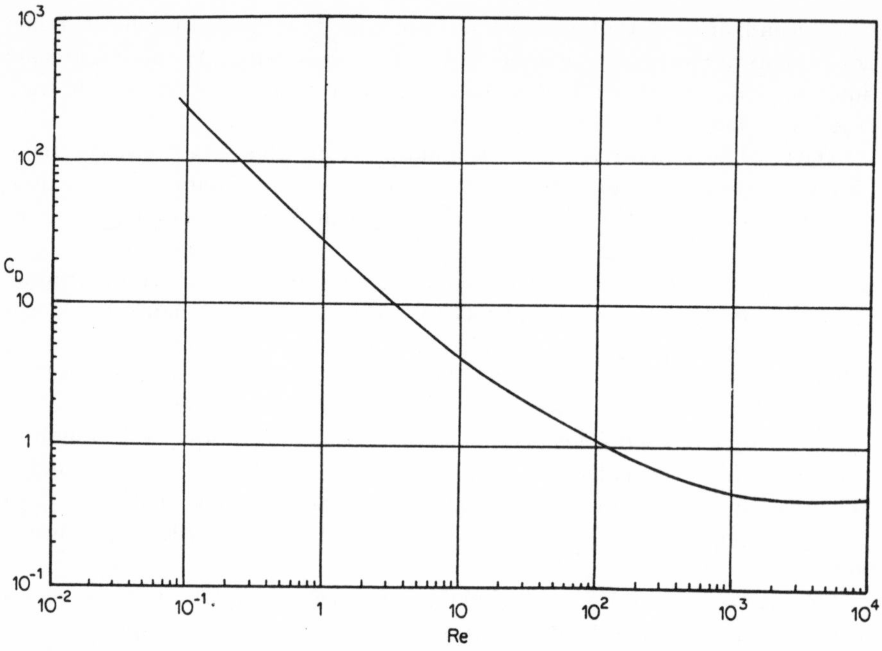

Figure 2.2. Relationship between drag coefficient and Reynolds number for a sphere.

Some of the different patterns of flow experienced by spherical water drops are illustrated in Fig. 2.3 and are discussed below. The change from one flow regime to another may be quite gradual and so the numerical values of Reynolds number which are quoted should be viewed as no more than approximate indications of regime boundaries. For Reynolds numbers in the range $10^{-5} < Re < 10^{-2}$ the flow is laminar and steady, and is symmetrical fore and aft of the sphere. Inertial forces are insignificant and the motion is dominated by viscous effects which pervade the whole flow field. With increase in Reynolds number, up to $Re \approx 20$, the fore-and-aft symmetry is progressively destroyed and, although the flow remains laminar and steady, the pressure at the front of the sphere becomes rather higher than that at the rear. For Reynolds numbers in the approximate range $20 < Re < 200$ the motion remains laminar and steady, but a further change in the flow pattern occurs.

At the rear of the sphere laminar separation takes place, and a standing eddy is created which increases in size with increasing Reynolds number; within this separated flow region a closed vortex ring is formed. In the Reynolds number range $200 < Re < 300$ the vortex ring rolls up, yielding a vortex loop which extends downstream to form a twin vortex trail. At this stage some lateral asymmetry of the flow pattern is possible with the consequential occurrence of a small side force on the sphere. For $300 < Re < 450$ the flow becomes increasingly unsteady, and vortex loops start detaching periodically from diametrically opposite sides of the sphere wake, producing a small but regular oscillatory side force on the sphere. With continued increase in Reynolds number the

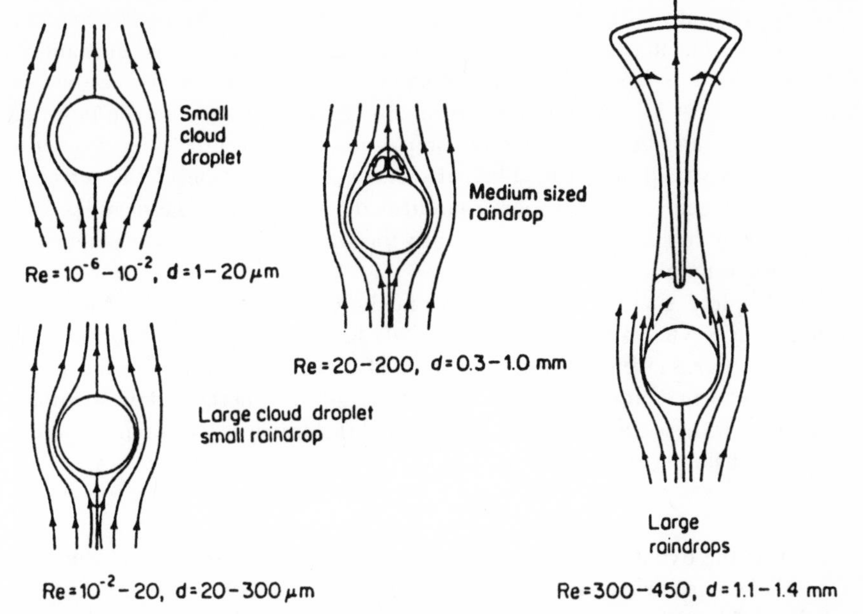

Figure 2.3. Air flow patterns about falling spheres as a function of Reynolds number. (Reproduced, by permission of Institute of Physics, from Mason, B. J. (1978), *Phys. Educ.*, **13**, 414–419.)

manifestation of viscous effects within the flow becomes progressively more confined to a narrow region on the surface of the sphere, the so-called boundary layer, and in the wake downstream of the sphere. The highest Reynolds numbers of interest in the aerodynamics of rainfall are about 4000. In the Reynolds number range $450 < Re < 4000$, and indeed well beyond this upper limit, the boundary layer flow over the surface of the sphere is laminar up to separation. Within the free shear layer defining the outer edge of the wake downstream of the sphere the flow remains laminar for some distance.

As the Reynolds number increases beyond 450 the frequency with which discrete parcels of vorticity roll up and are shed downstream increases, and the position within the shear layers where this process takes place moves progressively closer to the rear surface of the sphere. Increasingly the fine detail of the motion in the wake becomes irregular and chaotic, although overall, the wake motion continues to exhibit a pronounced periodicity.

It is a point of interest to note that over the range of Reynolds numbers for which asymmetry and unsteadiness in the wake are observed (from approximately 200 up to the maximum at which rain falls) a droplet descending through still air will experience a small side force. This force is capable of causing the raindrop to deviate laterally from its vertical descent path.

2.3 TERMINAL VELOCITY

In general, the size of a water drop, its terminal velocity, the Reynolds number defining its motion, and the drag coefficient are all closely interrelated.

At its terminal velocity the drag force acting on the falling droplet is just equal to its weight. The mass of the droplet is equal to the product of its volume and its density, which is approximately 10^3 kg m^{-3}. The weight is given by the product (mass $\times$ g), where g has the value of 9.81 m s^{-2}. So the computation of the weight, and hence the aerodynamic drag, of a spherical water droplet of given size is straightforward. The subsequent determination of the terminal velocity is, in general, rather more complicated, since, as we have seen, the flow pattern about the sphere, and consequently the drag coefficient, C_D, depend upon the Reynolds number, *Re*.

For droplets smaller than approximately 0.05 mm (2×10^{-2} in.) the motion is at Reynolds numbers substantially less than unity. In this range the drag of a sphere is given by Stokes' law, named after the great nineteenth-century physicist who first unravelled the laws of motion of spheres at low Reynolds numbers. For these conditions $C_D = 24/Re$ and the drag force, D, of the sphere is given by the simple expression

$$D = 3\pi\mu V d \tag{2.3}$$

where μ is the dynamic viscosity of the air, and V and d are, respectively, the velocity of the particle relative to the air and the diameter of the sphere. At these low Reynolds numbers, where viscous forces in the flow are much more important than inertial forces, the drag force is directly proportional to the dynamic viscosity, but is totally independent of the magnitude of the air density.

Since the weight of the spherical droplet (and hence the drag force) is proportional to d^3 it follows immediately from equation (2.3) that the terminal velocity, V_T, of a droplet subject to Stokes' law is proportional to the square of the diameter.

But only the finest droplets in clouds are subject to Stokes' law, and the droplets which make up the bulk of clouds, or are precipitated as drizzle or rain, fall at Reynolds numbers substantially above a value of 1. Under such circumstances both the density of the air and its viscosity to some extent influence the magnitude of the drag force, and there is no simple relation between the drag coefficient and the Reynolds number. Here we have to refer to the general curve (Schlichting, 1979) relating C_D and *Re* for flow about a sphere (Fig. 2.2).

To compute the terminal velocity, V_T, at any altitude a trial-and-error calculation is necessary, using the following relations. The drag force, D, is given by

$$D = \frac{1}{2}\rho V_T^2 \times \frac{\pi d^2}{4} \times C_D \tag{2.4}$$

The weight, W, is given by

$$W = \frac{\pi}{6} d^3 \times \sigma \times g \tag{2.5}$$

where ρ and σ are respectively the densities of air and water. Without serious error the density of water can be assumed constant at 10^3 kg m^{-3}, but the density of air varies substantially with altitude, and this effect must be taken into account if we are interested in calculating the terminal velocity at different altitudes.

At the terminal velocity, $W = D$; under this condition the drag coefficient, C_D, used in equation (2.4) must be consistent with the Reynolds number, Re, defined by

$$Re = \frac{\rho V d}{\mu} \tag{2.6}$$

where μ is the dynamic viscosity of the ambient air.

The above calculation procedure applies to droplets of spherical shape, i.e. for droplets with a diameter, d, less than about 1.5 mm.

For larger droplets, although their shape is no longer spherical, their frontal area remains circular, and their size can be specified by reference to the diameter of the frontal area. However, for a droplet of given volume, the deviation in shape from the spherical depends on the terminal velocity, which itself depends on the

Table 2.1 Velocity at which water droplets descend through the air at sea level (T = 20 °C; ν = 0.15 cm^2 s^{-1}).

Diameter (mm)		Velocity (m s^{-1})	Reynolds number
	0.001	3×10^{-5}	2×10^{-6}
	0.01	3×10^{-3}	2×10^{-3}
	0.02	0.012	1.6×10^{-2}
	0.05	0.073	0.24
Cloud	0.1	0.27	1.8
	0.2	0.72	9.6
	0.4	1.6	43
	0.6	2.5	100
Drizzle	0.8	3.3	180
	1	4.0	270
	1.2	4.6	370
	1.4	5.2	490
	1.6	5.7	610
	1.8	6.1	730
Rain	2	6.5	870
	2.2	6.9	1010
	2.4	7.3	1170
	2.6	7.6	1320
	2.8	7.8	1460
	3	8.1	1620
Break-up of	3.5	8.5	1980
raindrop	4	8.8	2350
possible	4.5	9.0	2700
	5	9.1	3030
	5.5	9.2	3370
Break-up certain	5.8	9.2	3560

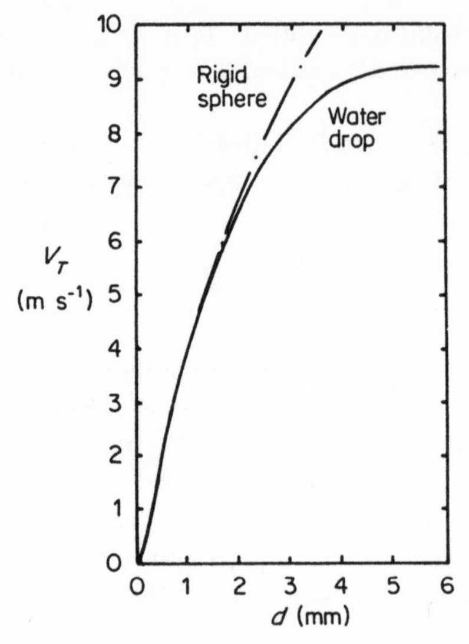

Figure 2.4. Comparison of the terminal velocities of raindrops and spheres (After Mason, B. J. (1978), *Phys. Educ.*, **13**, 414–419.)

altitude at which the droplet is falling. As a consequence the relationship between the size of the droplet, its terminal velocity, and the properties of the atmosphere becomes quite complicated.

Table 2.1 shows some representative data on the variation of terminal velocity with droplet diameter. The data have been computed using sea level conditions of pressure and an ambient temperature of 20 °C, for which the density, ρ, is $1.2\ \mathrm{kg\ m^{-3}}$, the kinematic viscosity, ν, is $1.5 \times 10^{-5}\ \mathrm{m^2\ s^{-1}}$, and the dynamic viscosity, μ, is $1.8 \times 10^{-5}\ \mathrm{kg\ m^{-1}\ s^{-1}}$.

A comparison of the terminal velocities of water droplets and of rigid spheres is given in Fig. 2.4.

2.4 THE MAXIMUM SIZE OF A RAINDROP

To understand the factors which determine the maximum size of a raindrop it is convenient to consider the collision of two large drops of stable geometry, which combine to form a still larger raindrop whose configuration is unstable. Following collision the enlarged drop will accelerate whilst its weight exceeds the aerodynamic drag.

The raindrop assumes an increasingly non-spherical shape as it descends with a velocity approaching the terminal velocity. The distortion takes the form of a progressive flattening, the horizontal equator increasing in diameter at the expense of the vertical separation of the poles, due to the pressure distribution imposed by aerodynamic forces, as discussed previously. A concave depression becomes evident on the underside of the raindrop (Mason, 1978). There comes a stage when this configuration becomes highly unstable and the drop suddenly weakens towards the centre and the liquid in that region expands rapidly upwards to form a sort of bubble supported on a toroidal ring of liquid (Fig. 2.5). Almost instantaneously, the bubble bursts to form a fine droplet spray and, in response to this disturbance of its own finely balanced equilibrium, the toroid itself fragments into a number of individual droplets. The precise conditions leading to disintegration of raindrops are still incompletely understood, but it appears that a crucial stage is reached when the surface curvature is diminished so that surface tension, upon which the integrity of the drop depends, is no longer capable of withstanding the combined effects of the differential pressure across the poles resulting from the external aerodynamic pressure distribution and the internal hydrostatic pressure. For the sizes at which raindrops disintegrate the Reynolds number is such that the motion of the drop through the air is accompanied by an oscillating wake, and this unsteadiness is communicated to the droplet itself, and no doubt acts as a perturbational input sufficient to trigger break-up.

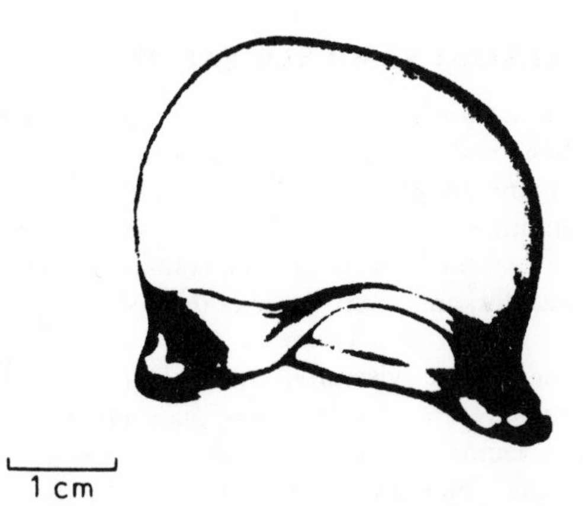

Figure 2.5. The incipient disintegration of a large raindrop. (Reproduced, by permission of Institute of Physics, from Mason, B. J. (1978), *Phys. Educ.*, **13**, 414–419.)

Chapter 3

The Descent of Ice Crystals, Hailstones, and Snow

3.1 THE AERODYNAMICS OF FALLING ICE CRYSTALS

Ice crystals occur in nature in two principal forms, both based on a hexagonal configuration. When crystal growth occurs principally on the prism faces, thin crystals known as plates are produced, whereas preferential growth on the basal faces results in columnar crystals. Related to these fundamental shapes, many other geometries occur: the capped column, columns terminated by pyramids, hollow columns, needles, and so on. Some of these shapes are illustrated in Fig. 3.1.

The shape of the crystal is primarily determined by the air temperature, although when the saturation level is less than 110 per cent the degree of saturation is also a factor. These considerations are summarized in Fig. 3.2, which illustrates the conditions under which various shapes grow. The approximate positions of regime boundaries are indicated. When a growing crystal moves from one region to another a composite crystal is formed. For example, a capped column (Fig. 3.1(c)) begins life as a simple column crystal. Descending through the air under gravity the crystal might pass from one region of Fig. 3.2 to another, and so the columnar growth stops and, instead, the basal faces extend outwards away from the longitudinal axis. So it is that all composite crystals are created by undergoing two or more distinct periods of growth.

Information on the aerodynamics of ice particles is important in a number of meteorological contexts. In particular a knowledge of the terminal velocity of crystals is required in calculating their rates of growth and in determining the contributions which particles of different sizes make to precipitation. Furthermore, there are relationships between the shape and size of an ice crystal and its orientation during descent; these factors have an important bearing upon the subject of meteorological optics, to which we shall return later in the chapter.

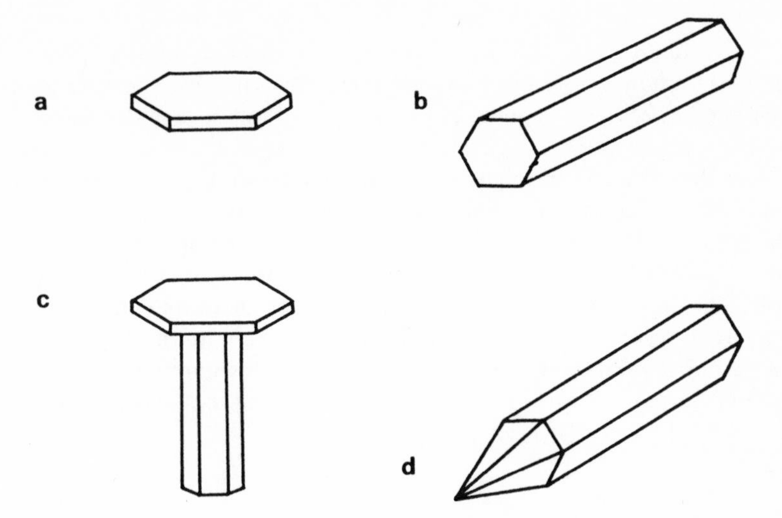

Figure 3.1. The basic geometries of ice crystals. a) plate; b) column; c) capped column; d) bullet.

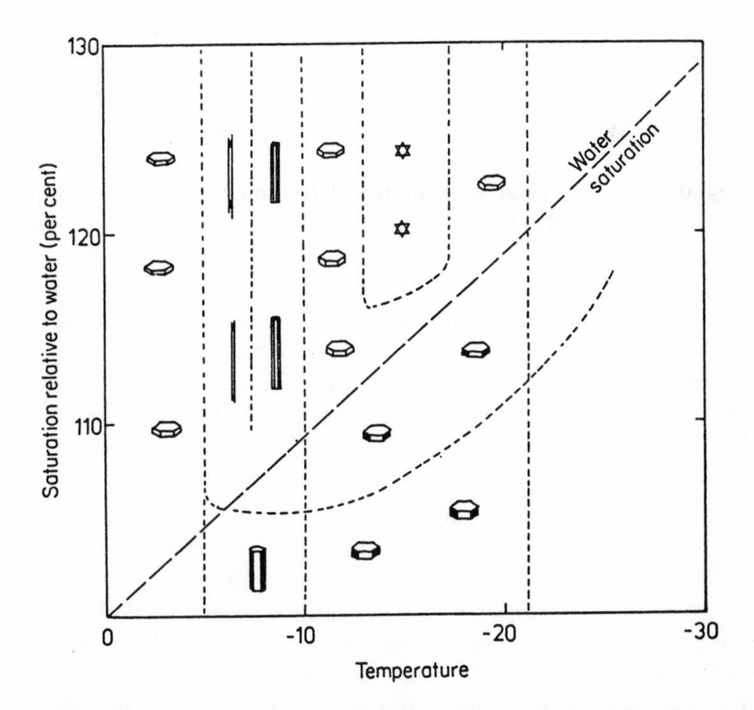

Figure 3.2. Relationship between ice crystal growth and ambient conditions.

The two principal crystal geometries of interest in meteorology are the plate and the column, and attention will be concentrated on the aerodynamic properties of these two forms. For convenience, we shall use the distance across the corners, w, of the hexagonal cross-section as a measure of the size of the ice crystal.

Under the influence of gravity these particles descend through the air and, as a consequence, experience aerodynamic forces. The balance between the weight of the ice crystal and the aerodynamic drag force acting upon it determines the terminal velocity of descent of the crystal. But the matter is not quite that simple. The drag force experienced by the crystal is affected by the orientation it adopts as it descends, and this is influenced by the pattern of air flow which passes around the particle. Once again, the Reynolds number is an important factor in determining the type of flow which the falling crystal experiences.

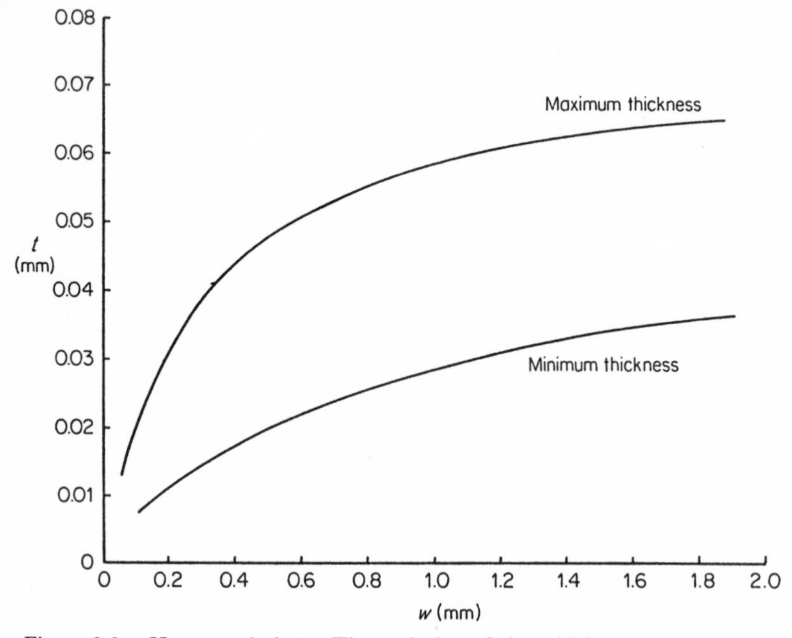

Figure 3.3. Hexagonal plates. The variation of plate thickness with distance across the corners.

In Fig. 3.3 the dimensions of typical hexagonal plate crystals are given. Maximum and minimum values of the plate thickness, t, are shown as a function of w. For small plates, $w \approx 0.1$ mm, the parameter t/w is typically about 0.2 whereas for larger plates, as a direct result of the preferential growth on the prismatic faces, there is a significant reduction in the ratio t/w with increase in w. These geometrical factors have a direct bearing on the weight of the individual crystals and, therefore, as a consequence of the equilibrium of forces at terminal velocity, upon the aerodynamic drag experienced by the particles.

The preferred direction of growth of columnar crystals is along the longitudinal axis and this is reflected in the fact that these crystals encompass a more restricted range of values of w than are found for plates. A representative range is 0.01 mm $< w <$ 0.3 mm, whilst the length, l, of these columns is such that the ratio l/w is typically in a range from 1 to 10.

Table 3.1 Summary of main properties of the motion of plate and columnar ice crystals.

(a) Plates ($t/w < 0.2$)

$w < 0.2$ mm	0.2 mm $< w <$ 4 mm	$w > 4$ mm
$V_T < 0.3$ m s^{-1}	0.3 m s^{-1} $< V_T <$ 1.5 m s^{-1}	$V_T > 1.5$ m s^{-1}
$Re < 1$	$1 < Re < 100$	$Re > 100$
$V_T \propto w \times t$	$V_T \propto \sqrt{t/C_D}$	—
Stable descent in any arbitrary orientation	Stable descent in orientation of maximum drag	Tumbling

(b) Columns ($1 < l/w < 10$)

$w < 0.1$ mm	0.1 mm $< w <$ 1 mm	$w < 1$ mm
$V_T < 0.7$ m s^{-1}	0.7 m s^{-1} $< V_T <$ 6 m s^{-1}	$V_T > 6$ m s^{-1}
$Re < 1$	$1 < Re < 100$	$Re > 100$
$V_T \propto w \times l$	$V_T \propto \sqrt{w/C_D}$	—
Stable descent in any arbitrary orientation	Stable descent in orientation of maximum drag	Unsteady motion

Note: Columnar crystals with $w > 0.3$ mm may be scarce or non-existent in nature.
Altitude 12,000 m (40,000 ft)
Assumed atmospheric properties:
Temperature = -15°C
Density = 0.25 kg m^{-3}
Dynamic viscosity = 1.63×10^5 kg m^{-1} s^{-1}

Whether columnar crystals with values of w greater than about 0.3 mm are able to maintain a separate existence is a matter of doubt, because, having developed terminal velocities in excess of 1 m s^{-1}, they are prone to collide with other particles, and to aggregate in the form of complex-shaped crystals or as hail. However, certain phenomena in meteorological optics provide indirect evidence for the occurrence in nature of large columnar crystals.

Results of tests (Clift *et al.*, 1978; Willmarth, 1964) on falling discs and cylinders, which are broadly similar in shape to plate and columnar crystals, respectively, provide valuable insights into the type of motion experienced by falling ice crystals.

In Figure 3.4 Reynolds numbers representative of the motions of plates and columns are given. Reynolds number is here defined by the expression

$$Re_w = \frac{wV_T}{\nu}$$

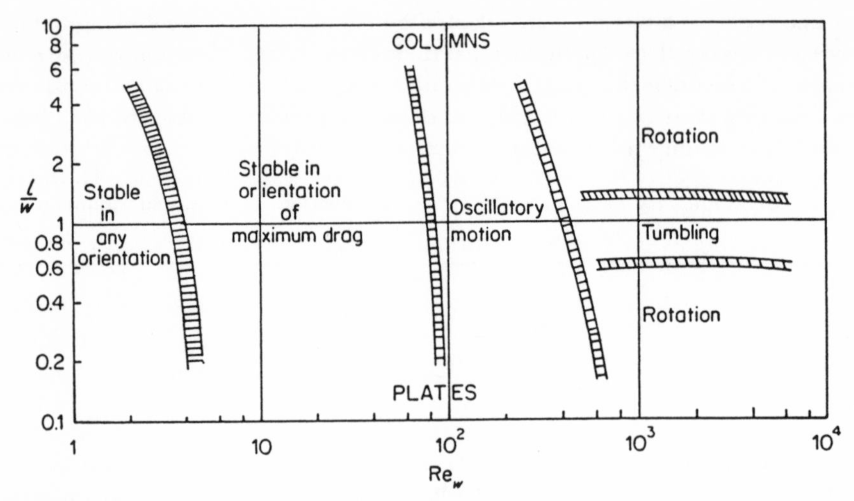

Figure 3.4. Regimes of motion of falling columns and plates. Plates have low values of l/w; columns large.

where V_T is the terminal velocity, w is the width across the corners and ν is the kinematic viscosity.

This diagram, when related to the aerodynamics of falling ice crystals, shows that they experience several distinct patterns of behaviour, and indicates the crucial importance of Reynolds number.

At low Reynolds numbers, typically less than about 1, particles only fall vertically and without rotation if they possess certain properties of symmetry or adopt specific stable orientations as they descend (Clift *et al.*, 1978). The idealized shapes of both the plate and columnar crystal possess three mutually perpendicular planes of symmetry and, as a consequence of their geometrical properties, in general, these particles have no preferred orientation. Once a plate or columnar crystal has adopted any arbitrary orientation it maintains that orientation as it descends further through the atmosphere. However, as an exception to the general rule, there is some evidence (Clift *et al.*, 1978) that long cylinders do tend to adopt a preferred orientation, with the longitudinal axis horizontal at these low Reynolds numbers.

For Reynolds numbers roughly in the range from 1 to rather less than 100, the ice crystal adopts a stable orientation corresponding to maximum drag. Thus a plate crystal descends with its principal axis vertical, whereas the longitudinal axis of a columnar crystal is maintained in a horizontal plane, as also are two flat surfaces of the prism. In this way both the plate and column maximize the projected area presented to the air.

At still higher Reynolds numbers, just below 100 or so, the ice crystals start to exhibit unsteady behaviour about the mean motion. A new parameter is important in this unsteady regime, a dimensionless moment of inertia, which depends upon the shape and size of the ice crystal and upon the ratio of the densities of the ice crystals and the surrounding air. Whether the unsteady motion of the ice crystal at these higher Reynolds numbers is in the form of small oscillations about a preferred orientation, or steady rotation about a specific axis, depends in a complex way on crystal shape, Reynolds number, and dimensionless moment of inertia. For a certain limited range of Reynolds numbers round 100 a plate crystal exhibits an oscillatory pitching motion as well as a substantial translational motion, floating to and fro rather like a leaf as it descends. However, the most representative type of motion adopted by plates as they descend at the highest Reynolds numbers is a continuous rotation about an axis joining two corners of the hexagon aligned in the horizontal plane.

Information relevant to columnar crystals, for $Re > 100$, is less clear. Under these conditions the crystal descends with a secondary motion superimposed upon the primary motion, the longitudinal axis, being maintained, on average, in the horizontal plane. The secondary motion initially probably takes the form of the longitudinal axis oscillating in a vertical plane about the mean orientation. At still higher Reynolds numbers there is some evidence (Clift *et al.*, 1978) that the cylinder rotates continuously about a vertical axis. For a narrow range of geometries, with $l/w \approx 1$, it seems that the crystal descends along a path inclined to the vertical, tumbling all the time in the direction of the trajectory.

Columnar and plate crystals for which Re_w is in the approximate range from 1 to 100 fall in such a way as to maintain a stable orientation. Any small disturbance to the orientation of the crystal gives rise to forces (and moments of forces) which tend to restore the orientation of the crystal to its initial position. The motion of the crystal is in stable equilibrium. For such conditions the flow in the immediate vicinity of the crystal and in its wake is laminar and steady, and any disturbances which might otherwise affect the orientation of the crystals are quickly damped out. At higher Reynolds numbers flow visualization experiments using circular cylinders show that regular oscillations of the flow patterns start to appear in the wake behind the body at Reynolds numbers of about 50. In the range $60 < Re < 5000$ vortices of opposite sense are shed alternately from opposite sides of the cylinder and form a regular pattern in the wake, which is commonly referred to as a Karman vortex street. These periodic disturbances within the flow pattern are communicated to the falling ice crystals, and this is the explanation of the oscillation or wobble which columnar crystals are known to display under certain conditions. The size of a typical crystal affected by this phenomenon is round about 1 mm. With size increasing from 1 mm up to 3 mm or more, there is a corresponding increase in Reynolds number and the frequency of vortex shedding in the wake also increases. The motion there becomes increasingly irregular and disordered and, eventually, at sufficiently high Reynolds numbers, the flow in the wake becomes turbulent.

Figure 1.5 on page 9 provides a useful impression of the different wake patterns behind a falling columnar crystal, as influenced by Reynolds number.

The wake pattern behind a plate crystal undergoes changes similar to those described for the column. For Reynolds numbers of about 100 there are two rows of staggered horseshoe-shaped vortex loops which systematically form on opposite sides of the laminar wake. At higher Reynolds numbers the wake pattern becomes increasingly disorganized and a fully developed turbulent wake is characteristic of high Reynolds number flows.

There is a close relationship between the aerodynamics of falling discs and hexagonal plates, and between circular cylinders and hexagonal columns. Figure 3.5 presents the drag coefficient, C_D, as a function of Reynolds number, *Re*, for discs and circular cylinders maintained in a fixed orientation with the maximum projected area transverse to the direction of motion. Figure 3.5 is based on the results of Roos and Willmarth (1971) for discs and Schlichting (1979) for circular cylinders.

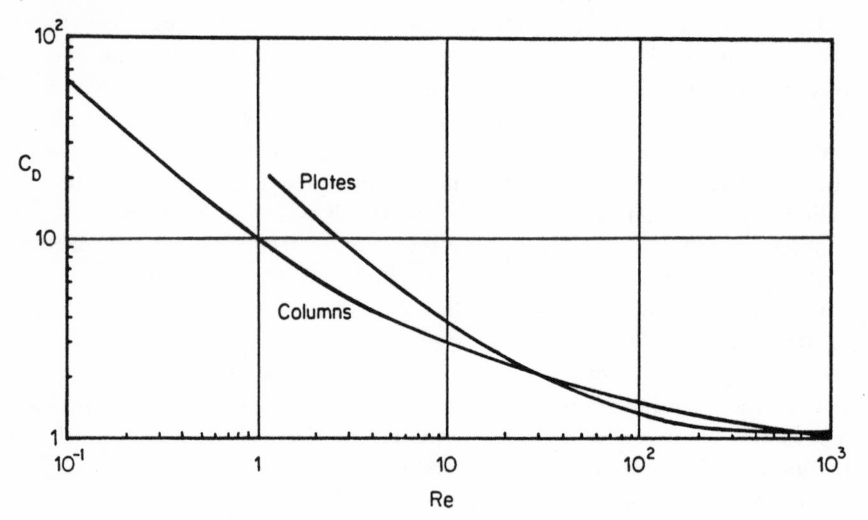

Figure 3.5. Relationship between drag coefficient and Reynolds number for columns and plates.

Here C_D is defined by the equation

$$D = \frac{1}{2} \rho V^2 S C_D \tag{3.1}$$

where ρ is the density of the air. In applying the data of Fig. 3.5 to ice crystals the reference area, S, is interpreted as follows. For columns S is defined by the product $w \times l$ and for plates S is taken as $3\sqrt{3}\ w^2/8$, which is the cross-sectional area of a hexagon.

3.2 TERMINAL VELOCITY OF FALLING ICE CRYSTALS

$1 < Re < 100$

It is convenient to start our calculation of terminal velocity of ice crystals by focusing attention upon the regime in which the crystals, whether they be plate or columnar, fall in a stable orientation with the maximum cross-section maintained in the horizontal plane. It is the larger crystals which are subject to the conditions of this regime, and because of the two distinct methods of growth, the two crystal types exhibit a divergence in geometry. Using the symbol h to represent the dimension of the crystal in the direction of the principal axis, we have $h = t$ for plates and $h = l$ for columns. Plates subject to this flow regime typically have values of h/w of the order 10^{-2}, whereas columns are represented by values of h/w of the order 10.

The weight of plate and columnar crystals is given by

$$W = \frac{3\sqrt{3}}{8} w^2 h \sigma g \qquad (3.2)$$

where σ is the density of ice of which the crystal is composed.

At the terminal velocity $W = D$, and so substitution of equations (3.1) and (3.2) in this relation yields an expression for the terminal velocity, V_T.

For plate crystals, noting $t = h$ we obtain

$$V_T^2 = \frac{2h\sigma g}{\rho C_D} \qquad (3.3)$$

an expression which is incidentally identical to that for a disc. Due to the preferential growth of plate crystals on the prismatic faces, the range of values of h is narrow and so for a given value of w, any variations of V_T due to the influence of h is relatively small.

For columnar crystals, with $h = l$, the terminal velocity is given by

$$V_T^2 = \frac{3\sqrt{3}}{4} \frac{w\sigma g}{\rho C_D} \qquad (3.4)$$

Apart from the possibility that C_D might be a weak function of aspect ratio h/w, this equation shows that the terminal velocity of columnar crystals in the regime under consideration is independent of l.

So, for plate crystals and columnar crystals falling under given atmospheric conditions we can present unique curves of terminal velocity as a function of the width across the corners only (Fig. 3.6).

Re $\ll 1$

At the opposite extreme, tiny columnar and plate crystals fall at low terminal velocities and have Reynolds numbers substantially less than unity. These tiny crystals have been subject only to small amounts of growth and do not display the extreme divergence in form of the larger crystals. For such small crystals the dimension h is of the same order of magnitude as w and, to a close approximation, the motion of these tiny crystals is comparable to that of a spherical particle of diameter $d = w$. For these low Reynolds number flow conditions the drag is given, by analogy with the flow about spheres, by

$$D = 3\pi\mu V w \qquad (3.5)$$

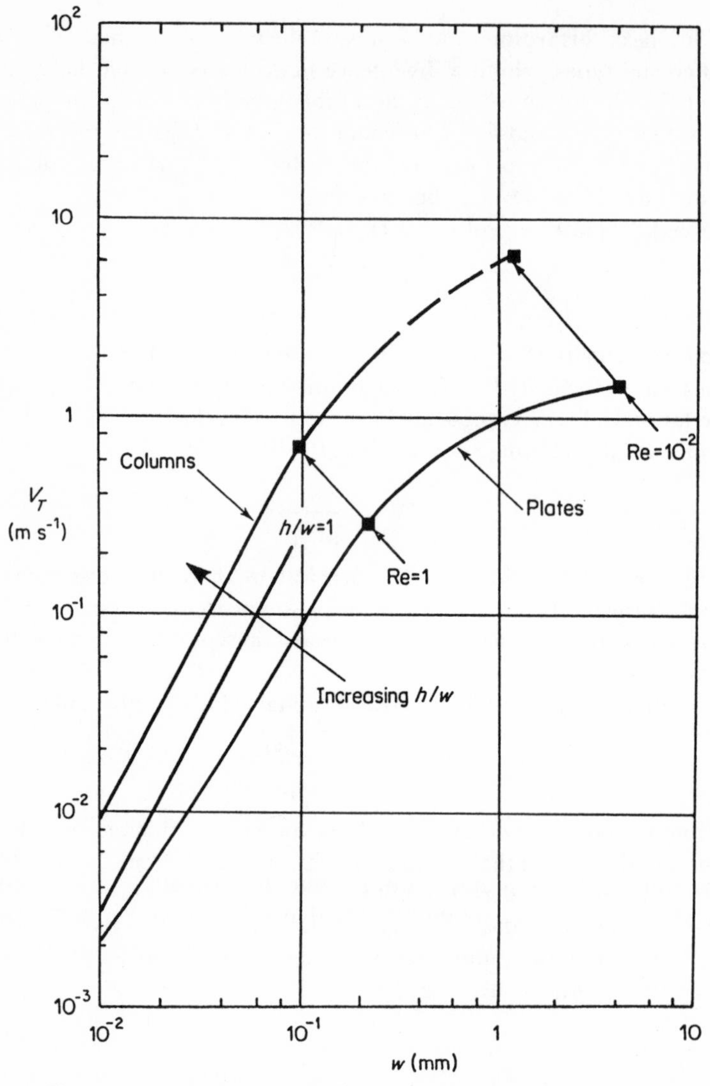

Figure 3.6. Terminal velocity as a function of crystal size for columns and plates.

and so, combining equations (3.2) and (3.5), the terminal velocity is given by

$$V_T = \frac{\sqrt{3}}{8\pi} \frac{\sigma g w h}{\mu} \tag{3.6}$$

which for $h/w = 1$ is

$$V_T = \frac{\sqrt{3}}{8\pi} \frac{\sigma g w^2}{\mu} \tag{3.7}$$

$10^{-3} < Re < 1$

Between the conditions corresponding to these tiny particles ($h/w = 1$) and the fully fledged plate crystals (h/w of order 10^{-2}) and columnar crystals (h/w of order 10) there are crystals in an intermediate state of growth, but subject to motion at Reynolds numbers less than unity. The orientation of these crystals during descent is arbitrary and this must affect the terminal velocity. However we can obtain some insight into the approximate magnitude of V_T by recasting equation (3.6) in the form

$$V_T \approx \text{const} \times \frac{\sigma g w^2}{\mu} \times \left(\frac{h}{w}\right) \tag{3.8}$$

an expression which shows that V_T increases approximately as the square of the size of the particle.

The important atmospheric conditions for the formation of ice crystals in temperate climates are in the ranges: (1) altitudes between 9000 m (30,000 ft) and 15,000 m (50,000 ft); (2) temperatures from −5 °C to −25 °C. We shall make some sample calculations at an altitude of 12,000 m (40,000 ft) assuming an air temperature of −15 °C. At this temperature the dynamic viscosity, μ, of air is 1.63×10^{-5} kg m^{-1} s^{-1} and the density of air, at $t = -15$ °C and a pressure of 1.85×10^4 N m^{-2}, is 0.25 kg m^{-3}. A broad summary of the calculations is contained in Table 3.1.

Figure 3.6 shows the variation of terminal velocity with size of ice crystal for typical plate and column crystals.

Comparing plate and columnar crystals with equal widths across the corners, the columnar form is seen to have the greater terminal velocity. This comes about simply because the columnar crystal is the heavier, by a large margin, and this is the dominant factor in determining the speed of descent.

Plate crystals with $w < 0.2$ mm and columnar crystals with $w < 0.1$ mm descend at Reynolds numbers less than 1. Under such conditions the aerodynamic drag is almost entirely due to viscosity, and is essentially independent of the density of the air. The dynamic viscosity, μ, of air varies only slowly with temperature and is unaffected by variations in atmospheric pressure. Consequently, for columnar and plate crystals in this specific size range, the terminal velocity depends only on the shape, size, and orientation of the crystal, and is insensitive to altitude. Consequently, the data in Fig. 3.6 for $Re < 1$ provide a useful representation of the relationship between terminal velocity and particle size anywhere in the earth's atmosphere. Furthermore the densities of water and ice are similar (1000 kg m^{-3} and 920 kg m^{-3} respectively) so that ice crystals with $l/w \approx 1$ and $w < 0.1$ mm fall at virtually the same velocity as raindrops of the same size.

Larger crystals fall at higher Reynolds numbers, and the effects of inertial forces, which depend upon air density, assume increasing importance. The density of the air at the representative conditions on which our calculations are based is less than a quarter of the sea level value. In consequence since, at terminal velocity, the aerodynamic drag and the weight of the crystal are in balance, and the weight depends only upon the size of the crystal, in the more rarefied conditions at high altitude the terminal velocity of the larger crystals is well above the corresponding sea level value.

3.3 OPTICAL PHENOMENA IN THE ATMOSPHERE

Ice crystals are capable of acting as optical prisms, influencing the path followed by a ray of light from the sun to the human eye. The interrelationship between sunlight and crystals of ice suspended in the atmosphere is responsible for a range of spectacular optical effects. These are known collectively as haloes, although the halo is just one of a range of displays whose components, arcs of light and colour, may be seen collectively or individually. Here we shall briefly consider the scientific basis of these beautiful effects.

The vital influence of aerodynamic factors on the orientation of the crystals has already been established. The angles between the faces of the important crystals in meteorological optics satisfy the following conditions: (1) between adjacent prism faces the angle is 120°; (2) between alternate prism faces the angle is 60°; (3) the junctions between the basal and side faces form an angle of 90°. It is these angular combinations which are responsible, at least in part, for the optical phenomena which we shall consider in greater detail here.

The most important phenomena of optical meteorology resulting from ice crystals are illustrated diagramatically in Figure 3.7. Phenomena caused by the refraction of light can often be seen in dazzling colours; those whose origin is reflected light are usually colourless.

The 22-degree halo, the commonest of the atmospheric haloes, is formed by sunlight passing through alternate side faces of randomly orientated ice crystals. It is observed as a ring of light of varying intensity surrounding the sun.

When the light is refracted through a basal face and a side face of randomly orientated plate or columnar crystals, the 46-degree halo results.

The parhelia, known as 'mock suns', are the commonest optical phenomena caused by oriented crystals, and the shapes responsible are plates (and possibly capped columns and bullets) descending with their longitudinal axes aligned with the vertical direction. The parhelia, visible as bright spots of light at the same elevation as the sun and just outside the boundaries of the 22 degree and 46 degree haloes, are caused by light passing through alternate side faces of the crystals. As they are caused by refracted light, the colours of parhelia are often quite beautiful. These same crystals are the source of the circumzenith arc, which is much rarer than the parhelia because it is only formed when the elevation of the sun is less than 32.2 degrees. It is even more rarely observed because people do not often look directly above themselves into the sky, which is the direction in which the arc lies. The circumzenith arc is formed by light entering the upper horizontal face of a crystal and emerging through a vertical side face. These same crystals are responsible for the complementary phenomenon, the circumhorizon arc, which is formed by light entering a vertical side face and leaving through the bottom horizontal face. The circumhorizon arc only appears in the sky when the elevation or the sun is greater than 57.8 degrees. As this arc is only formed when the sun is high in the sky, it is one of the few atmospheric haloes that is restricted to certain areas of the earth's surface, being confined to latitudes from 55.7 degrees north to

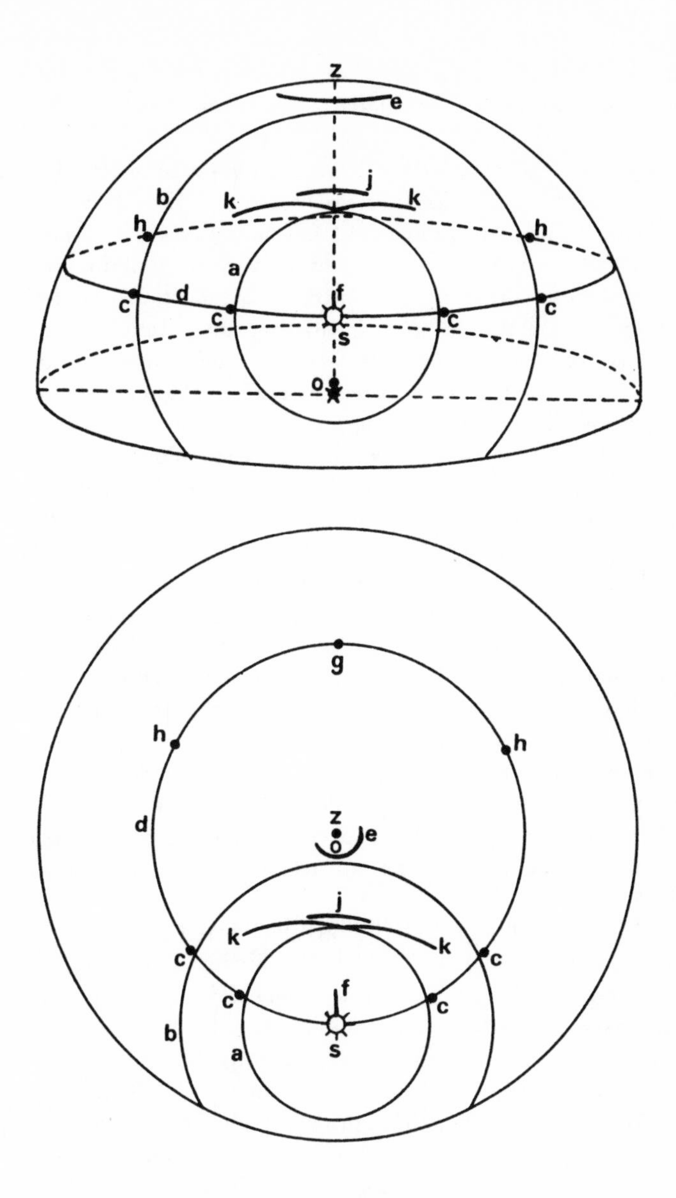

Figure 3.7. Diagrammatic representation of some of the phenomena of meteorological optics. The upper diagram presents a projection, in the plane of the sun and observer, from a point outside the hemisphere. The lower diagram presents a plan view with the observer beneath the zenith. **s**, Sun; **o**, observer; **z**, zenith; **a**, 22-degree halo; **b**, 46-degree halo; **c**, parhelia; **d**, parhelic circle; **e**, circumzenith arc; **f**, solar pillar; **g**, anthelion; **h**, paranthelion; **j**, upper Parry arc; **k**, circumscribed halo.

55.7 degrees south. Furthermore, ice-bearing cirrus clouds are scarcer in summer, when the sun's elevation is at its highest, so this phenomenon will always be a rarity.

The parhelic circle, visible in the sky as a circle at the same elevation as the sun, depends upon external reflection from orientated crystals—the vertical side faces of capped columns and plates, with their longitudinal axes in the vertical direction, and from the end faces of columns aligned in the horizontal direction. A related phenomenon, the solar pillar, results from external reflection from the basal face of plates and capped columns, descending with the principal axis vertical, or from the horizontal faces of columnar crystals with the longitudinal axis aligned in the horizontal plane. This phenomenon, seen as a pillar of light above the sun, is most often visible above the rising or setting sun.

Many people spend their entire lives without seeing a halo display. Yet they are not uncommon phenomena. For example, it has been estimated (Greenler *et al.*, 1979) that in Wisconsin, USA, some or all of the elements of the 22 degree halo complex are visible, on average, about 160 days in each year. Again, over southern England observers report (Goldie *et al.*, 1976) that halo elements can be seen, on average, about one day in three. Of course, only meteorologists and observers with a special interest in haloes will see them this frequently. My own experience may be closer to that of the typical reader. Since my interest in this topic was aroused a few years ago, I have regularly scanned the skies. I find that one quite often sees small patches of light in the sky which one can deduce are probably elements of a halo display; such events I personally regard as of general interest but having no special merit. On the other hand, when a well-defined halo display covers a substantial portion of the sky it is something very special—a marvellous array of colours, and quite awe-inspiring. An event of this nature I personally witness on only about ten days of the year.

I hope that as a result of the descriptions given here I can persuade readers not previously acquainted with haloes to look out for them in the future. Their patience will surely be rewarded, for they are fascinating and beautiful phenomena.

Over the years a great deal of information about haloes has been collected as a result of direct observations of the sky. But recently the availability of computers has provided a new and powerful means of studying halo phenomena. The path of a ray of light from the sun passing through an individual ice crystal can be simulated by the computer and, by considering the collective effect of a multitude of crystals of appropriate orientation distributed throughout the atmosphere, a wide variety of haloes can be modelled. Computer simulations of this kind have been published by Greenler (1980). Not only have these computer calculations reproduced many of the well-known haloes, but they have also proved capable of confirming certain halo phenomena which have only been sighted on a few occasions. Indeed these calculations can predict the existence of faint effects which may never have been recognized for what they are, indicating at the same time the part of the sky to which observation should be directed.

In conclusion we note that whenever a halo display is visible in the sky it tells us something of the size and terminal velocity of the ice crystals present in the atmosphere; for, as we have seen, the various optical phenomena are caused by crystals possessing particular orientations, and these orientations depend in turn upon the Reynolds number and, ultimately, on the size and terminal velocity of the ice particles, as described in Table 3.1.

3.4 HAILSTONES

Some parts of the world are much more prone to the damaging effects of hailstones than others. Parts of China, India, South Africa, Argentina, central Europe, the western plains of the United States, and the Caucasus region of Russia are areas which have the highest frequencies of hailstorms. Hailstones come in a wide range of sizes. A typical hailstone is perhaps 0.5 cm (0.2 in.) in diameter, but in regions where hailstorms are frequent, hailstones as large as 2 cm (0.8 in.) in diameter are not uncommon. In exceptional circumstances even larger stones are formed. The largest reliably recorded hailstone was irregular in shape, but was approximately 14 cm (5.5 in.) in diameter. It fell on Coffeyville, Kansas, USA, during September 1970 and weighed 766 g (1.67 lb).

The origins of hailstones can be traced to the appearance of tiny ice particles amongst the supercooled cloud droplets of cumulonimbus clouds. The growth of the ice particles involves the complex interaction of meteorological and aerodynamic factors. Initial growth of the embryo hailstones is by condensation but once the particles have reached a certain size, round about 0.1 mm in diameter, subsequent growth is primarily as a result of collisions with other particles, consequent upon their different terminal velocities. Strong updraughts are a feature of cumulonimbus clouds, so that an embryo hailstone is swept to higher altitudes whenever it encounters rising air with an updraught velocity in excess of its terminal velocity. Sooner or later the updraught releases the stone and again it falls, collecting more water. This process may be repeated on several occasions, before eventually the terminal velocity of the hailstone, which increases as the stone grows in size, is sufficient to ensure that it falls to the ground.

Due to the complicated nature of the processes of formation, the mean density, shape, surface texture, and size of a hailstone all depend upon the particular circumstances surrounding its growth pattern from its inception through to its final state. The following broad characteristics can be identified but, because of the wide variety of conditions under which hailstones grow, exceptions to the general description presented here inevitably exist.

The smallest hailstones are in the form of hail pellets, typically having a maximum dimension of about 1 mm. Such hail may have a density as low as 0.1 g cm^{-3}. Hail of this type consists of a soft, friable, open structure, resulting from the growth of the pellet by a process of rapid, localized freezing of individual droplets following impact, and substantial volumes of air are entrapped within the frozen structure. This type of hail is often referred to as soft hail or graupel. With increasing size there is an increasing tendency for the hailstone growth to take place with wet conditions on the surface. This results in a more uniform distribution of the accreted water over the surface of the hailstone and less air is entrapped. Consequently there is an increase in density, so that a hailstone of 0.5 cm diameter may have a density typically in the range from 0.7 to 0.9 g cm^{-3}. Hailstones with a diameter of 1 cm or more have a density essentially that of pure ice, which is 0.92 g cm^{-3}.

Just as the average density of a hailstone is broadly related to its size, so also is its shape. Soft hail pellets, having a maximum dimension of a few millimetres, are generally conical in shape with a rounded base. Such hail pellets fall to the ground with the rounded base leading, and it is on this surface that accretion mainly takes place. With increase in size there is a tendency for the hailstone to lose its preferred orientation. As it moves through the atmosphere it acquires a general tumbling motion resulting in a more uniform level of accretion over its entire surface. This is the process of formation leading to true hailstones, which are essentially spherical in shape and typically are perhaps 0.5 cm in diameter.

Large hailstones up to between 2 and 4 cm retain a spherical shape, but as the size increases a greater proportion have the shape of oblate spheroids. The appearance of giant hailstones is different again. A typical giant hailstone is roughly spherical in shape, or approximates to an oblate spheroid, but substantial rounded lobes or knobbly protuberances which extend outwards from the nominal profile give it an irregular surface configuration.

Terminal velocity

The terminal velocity of a hailstone is computed in the following way, assuming it may be represented by a corresponding equivalent sphere.

The weight, W, of the hailstone is given by

$$W = \frac{\pi d^3}{6} g\sigma \tag{3.9}$$

where σ is the mean density of the hailstone, d is its nominal diameter, and g is the gravitational acceleration.

The aerodynamic drag, D, experienced by the falling hailstone when it reaches terminal velocity, V_T, is

$$D = \frac{1}{2} \rho V_T^2 \times \frac{\pi d^2}{4} \times C_D \tag{3.10}$$

where ρ is the air density and C_D is the drag coefficient. At the terminal velocity the weight and drag are in equilibrium, so that equations (3.9) and (3.10) can be combined to give

$$V_T = \left(\frac{4}{3} \frac{\sigma}{\rho} \frac{gd}{C_D}\right)^{1/2} \tag{3.11}$$

The magnitude of the drag coefficient in equation (3.11) is affected by the shape, orientation, and surface texture of the hailstone, and by the Reynolds number, *Re*, given by

$$Re = \frac{Vd}{\nu} \tag{3.12}$$

where ν is the kinematic viscosity of the air.

For a given diameter, d, the main uncertainty in computing the terminal velocity using equation (3.11) concerns the magnitude of C_D.

Aerodynamics

The drag characteristics of smooth spheres and spheres with regular surface roughness elements, achieved by covering the surface of the sphere with uniform tiny glass beads, have been thoroughly researched, and their fluid mechanics is well understood (Achenbach, 1974; Schlichting, 1979). Smooth spheres exhibit a dramatic fall in drag coefficient, as Reynolds number is increased, at a critical Reynolds number of the order of 3×10^5. The sudden change from subcritical to supercritical flow about the sphere is associated with the nature of the boundary layer on the surface of the sphere. In subcritical flow separation of a laminar

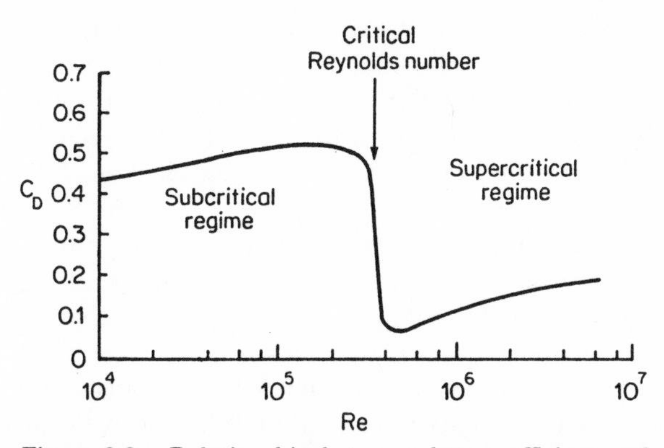

Figure 3.8. Relationship between drag coefficient and Reynolds number for a sphere with a smooth surface. This diagram is consistent with Fig. 3.2 and is an extension of that diagram to higher Reynolds numbers.

boundary layer leads to a broad wake and a high drag coefficient. In supercritical flow the boundary layer at separation is turbulent. This layer is better able to withstand the adverse pressure gradient at the rear of the sphere, separation is postponed, the wake which is ultimately formed is narrower than for laminar flow, and this accounts for the very considerable reduction in drag coefficient.

The effect of surface roughness is to promote an earlier transition from laminar to turbulent flow on the surface of the sphere, with the result that the critical Reynolds number decreases with increase in relative roughness, k/d, where k is the height of the roughness element. This trend is exhibited for values of k/d up to about 0.0125, the highest value tested (Achenbach, 1974) with this type of roughness.

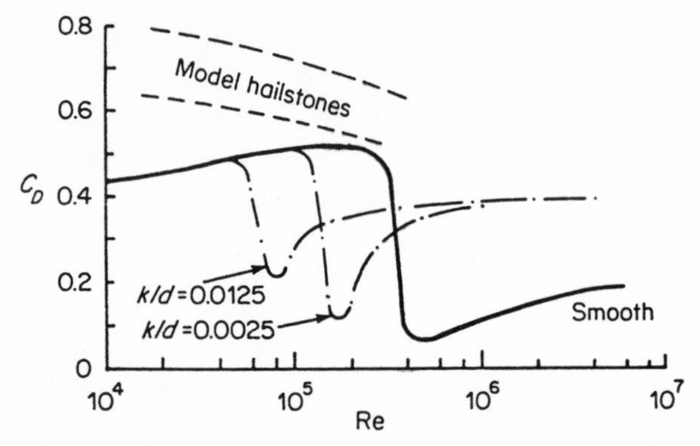

Figure 3.9. Relationship between drag coefficient and Reynolds number. Data are given for a smooth sphere and spheres with artificially roughened surfaces. Also shown are test results for models of hailstones with large surface protuberances.

The surface roughness of giant hailstones is highly irregular and different in character from the regular roughness elements previously discussed. Furthermore, the height of the irregular roughness on the surface of a hailstone is generally in excess of $0.1d$. Tests on artificial hailstones, extending up to $Re = 2 \times 10^5$ or more, have been reported (Bailey and Macklin, 1968; Macklin and Ludlam, 1961).

Evidence of a critical Reynolds number was found for hailstones of 4 and 6 cm diameter. However, for larger hailstones, up to 10 cm diameter, no dramatic fall in drag coefficient was evident. It seems that the gross surface imperfections of giant hailstones suppress this particular effect, and the drag coefficient exhibits a slow, regular variation with Reynolds number.

A range of drag coefficients from 0.5 to 0.8 has been found for the largest artificial hailstones at high Reynolds numbers. Investigations of their internal structure suggest that giant hailstones tumble as they descend, so that the projected area of the hailstone constantly changes. Taking all of the above evidence into account a mean value of $C_D = 0.65$ would seem to be representative of conditions experienced by giant hailstones. Since $V_T \propto C_D^{-1/2}$, the uncertainty in the estimation of V_T is of the order $\pm$ 10 per cent.

In order to estimate the approximate descent speed that hailstones attain at ground level the following representative values will be used:

$$\sigma = 920 \text{ kg m}^{-3}, \rho = 1.21 \text{ kg m}^{-3}, g = 9.81 \text{ m s}^{-2}, \nu = 1.5 \times 10^{-5} \text{ m}^2 \text{ s}^{-1}.$$

Assuming a drag coefficient of $C_D = 0.65$, the results in Table 3.2 are obtained from equation (3.11).

Using these numerical values in equation (3.11), which can be recast in terms of the mass m of the hailstone in the form

$$V_T = \left(\frac{4}{3}\frac{\sigma g}{\rho C_D}\right)^{1/2}\left(\frac{6m}{\pi\delta}\right)^{1/6}, \tag{3.13}$$

the terminal velocity of the hailstone is given to a good approximation by

$$V_T = 14\, m^{1/6} \tag{3.14}$$

where V_T is in m s^{-1} and m is in grams.

Table 3.2 Characteristics of hailstones at sea level (C_D = 0.65).

d(cm)	V (m s^{-1})	m (g)	Re
0.5	8.65	0.06	2.9×10^3
1	12.2	0.47	8.1×10^3
2	17.3	3.8	2.3×10^4
5	27.3	58.9	9.1×10^4
10	38.7	471	2.6×10^5
20	54.7	3770	7.3×10^5
30	67	12,700	1.34×10^6

From time to time unconfirmed reports emerge, particularly from India and China, of hailstones of truly gigantic proportions. The following news item, for example, sent by their correspondent in Peking, appeared in the London *Daily Telegraph* of Wednesday 15 April 1981:

> KILLER HAILSTONES. Hailstones weighing up to 30 pounds (14 kg) each slammed into an area of Southern China recently killing five people and injuring 225, the *Canton Evening News* reported. The paper said the storm destroyed or seriously damaged more than 10,500 homes.

A number of similar accounts of giant stones have been collected together by Ludlam (1980) in his book *Clouds and Storms*, though none of these references matches the size of the hailstones mentioned above.

The calculations in Table 3.2 give some interesting insights into the newspaper item concerning the Chinese hailstones. A hailstone of 30 lb (14 kg) would have had an effective diameter of just over 30 cm (comfortably exceeding the 22 cm of an ordinary football) and a terminal velocity of about 70 m s^{-1} (or 157 miles per hour). The kinetic energy possessed by such a hailstone would be about 34,000 J. Only under quite exceptional circumstances could this energy be absorbed and the hailstone remain intact after descent. Direct impact on the ground would inevitably result in the hailstone shattering. If they do nothing else, these calculations indicate that the claimed size of these giant hailstones should be treated with considerable caution.

Large blocks of ice have indeed been collected after violent thunderstorms in China, but scientists who have had the opportunity to inspect these specimens have concluded that they were not hailstones and had been formed by some other process. The smooth external surface and the lack of a seed nucleus were two factors which pointed to alternative mechanisms of formation.

3.5 SNOW

The commonest forms of snowflake are aggregates of ice crystals, though an examination of some snowflakes shows them to be large forms of ice crystal. Ice crystals and supercooled water droplets are often brought into coexistence within a cloud, at sub-freezing temperatures. This is an unstable state and immediately the ice crystals grow rapidly in size at the expense of the water droplets, which evaporate. As they grow, the crystals start to descend under the influence of gravity. On their way down the crystals collide with supercooled water droplets and with other ice crystals and in this way snowflakes are formed. If the air temperatures are sufficiently low the snowflakes retain their identity down to ground level. However if they encounter warmer air during their descent the snowflakes melt and precipitation occurs in the form of rain or sleet. It is a matter of observation that, in general, snowflakes fall much more gently to the ground than rain. The lower terminal velocity of snowflakes is explained by the fact that, whereas the mass of raindrops is concentrated into a given volume, a snowflake of given mass presents a very large projected area to the air through which it descends. It is not possible to provide any very precise figures on the terminal velocities of snowflakes due to the infinite variety of shapes that are found to exist, but 1–2 m s^{-1} is a typical figure (Hobbs, 1974).

Chapter 4

The Airborne Dispersal of Fruits and Seeds

Nature has evolved a variety of mechanisms to ensure that one generation of plant life is succeeded by another. Some plants, which die back to ground level in winter, contain a store of food which is used to produce new horizontal underground stems, from which upward growth develops in the spring. An example is the garden mint (*Mentha viridis*). Other plants use a rather similar process, but the main activity takes place above ground level. Such plants as the strawberry (*Fragaria vesca*) produce runners from which new plants grow. These new plants take on an independent existence when the parent plant dies or the runner withers and becomes severed from the parent. These are examples of vegetative asexual reproduction. Many plants, however, have evolved sexual reproduction systems, the end result of which is the formation of a seed or seeds, from which the new, independent, life will ultimately spring. This form of reproduction is centred upon the flowers of the parent stock. Within a flower the seed develops from the ovule, after fertilization has taken place through pollination. At maturity a seed consists of an embryo, which is surrounded by a thin, dried layer of tissue, known as the inner integument, which is itself enveloped within an external seed coat, the outer integument or testa. The testa has to serve a number of functions, including that of protecting the embryo and, in some cases, it has an important role in aiding dispersal. In many flowering plants, besides the formation of seeds, other important changes take place after fertilization. The fruit of a plant is formed as a result of developments to the ovary and other associated parts of the flower. In the most elementary case the fruit consists simply of the mature ovary, plus the seed which it surrounds. More generally, however, the fruit can be quite a complex structure, consisting of one or more mature ovaries interconnected by other components derived from floral parts, the whole assemblage serving to protect and subsequently to facilitate the dispersal of the seed or seeds contained within.

Disseminules—that is fruits and seeds—have evolved various means of dispersal. Here our interest lies with those plants which secure the movement of their seeds or fruit from the parent by exploiting aerodynamic principles. A number of distinctive flight styles are to be found (Burrows, 1975) and it is convenient to deal with these separately.

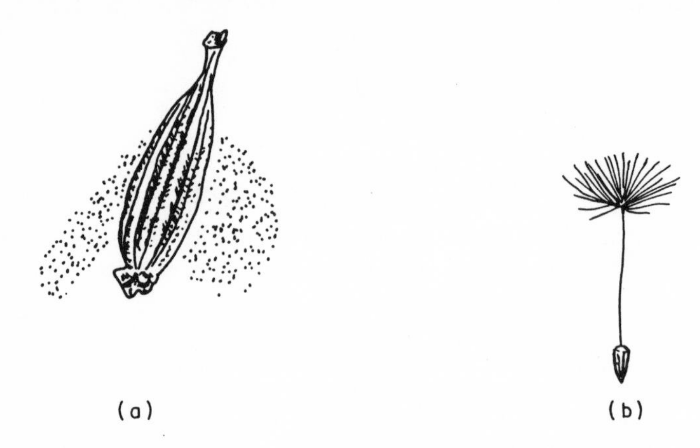

Figure 4.1. Airborne dispersal by means of aerodynamic drag: (a) capsule of an orchid (*Cymbidium*), showing the tiny seeds being dispersed by the wind; (b) the parachute of the dandelion (*Taraxacum*).

4.1 DUST SEEDS, SPORES, AND POLLEN

The seeds of some plants, such as orchids (*Orchidaceae*), are tiny particles scarcely visible to the naked eye. So also are the spores produced by the more primitive groups of plants, such as fungi and ferns. Although variable in size and shape, many of these seeds and spores share the common properties of being very small, and compact in form, more or less spherical in shape, they are very light, and in consequence have a very low terminal velocity. These particles develop no significant lift and their airborne motion is dominated by viscous drag forces. In practice, conditions are rarely still enough for these particles to fall vertically; the lightest breeze or the existence of local convection currents is sufficient to ensure that they follow a complex trajectory from the parent to their final resting place. Thermals are localized regions within the atmosphere in which substantial upward movement of air is generated from ground level as a result of differential heating of the surface of the earth by the sun. Consequently, updraughts at velocities in excess of the terminal velocities of these tiny particles are a common feature of diurnal convection patterns; as a result the spores and seeds can be swept to high altitudes and can subsequently be carried horizontally over vast distances. Because of their mode of transportation, the spores and seeds in this group can be liberated close to the ground yet still achieve substantial ranges of dispersal.

These small particles operate at Reynolds numbers of order 1 or below, and their motion is in accordance with Stokes' law of resistance. In this regime, the drag, D, of the particle is directly proportional (1) to the relative velocity between the particle and the airstream, V, and (2) to the dynamic viscosity, μ, of the air.

For a spherical particle of diameter d, the resistance law is given by

$$D = 3\pi\mu V d \tag{4.1}$$

where in still air, V represents the terminal velocity. Many of the disseminules included in this group are roughly spherical in shape, so that equation (4.1) is a useful approximation to the drag of such particles, where d then represents a typical cross-sectional dimension, analogous to the diameter of a sphere.

It is also worth observing in passing that airborne pollen grains are in the same size range as dust seeds and spores, and consequently are subject to similar types of motion.

Table 4.1 Typical data for representative dust seeds, spores and pollen.

Name	Representative diameter, d (mm)	Mass, m (g)	Terminal velocity at sea level, V (cm s^{-1})	Reynolds number Re
Small fungus spore	0.005	5×10^{-11}	0.06	2×10^{-4}
Pollen of horse chestnut	0.015	1.4×10^{-9}	0.5	5.4×10^{-3}
Pollen of oak (*Quercus robur*)	0.036	2×10^{-8}	3	0.07
Seed of orchid (Orchidaceae)	0.1	5×10^{-7}	27	1.8

Note: The terminal velocity of fungus spores is typically in the range $0.05 < V(\text{cm s}^{-1}) < 3$. Several trees have pollen grains with terminal velocities close to that of the oak. Examples are pine (2.5 cm s^{-1}), willow (2.2 cm s^{-1}), elm (3.2 cm s^{-1}), lime (3.2 cm s^{-1}) and hazel (2.5 cm s^{-1}). All the above particles are subject to Stokes' flow for which the following relations apply, assuming the particles are approximately spherical and of similar densities: $m \propto d^3$, $V \propto d^2$, $Re \propto d^3$. For seeds larger than those of the orchid the simple relationships of Stokes' law are no longer valid.

At the terminal velocity, the weight of the particle

$$W = mg = \frac{\pi}{6} d^3 g \sigma \tag{4.2}$$

and the drag force, D, as determined by equation (4.1), are in balance so that the terminal velocity V_T is given by

$$V_T = \frac{1}{18} \frac{\sigma g}{\mu} d^2 \tag{4.3}$$

Alternatively, V_T can be expressed in terms of the mass, m, of the particle. Thus

$$V_T = \frac{1}{3\pi} \frac{g}{\mu} \left(\frac{\pi\sigma}{6}\right)^{1/3} m^{2/3} \tag{4.4}$$

What experimental evidence there is suggests that the density of dust seeds and spores is close to that of water soon after liberation, though as a result of desiccation there is some reduction of density with time.

In order to determine the approximate magnitude of the terminal velocity of particles at sea level the following values can be substituted: σ = 1000 kg m^{-3}, g = 9.81 m s^{-2}, μ = 1.8 × 10^{-5} N s m^{-2}. Using these data, from equation (4.3) the result

$$V_T = 30.3\ d^2 \tag{4.5}$$

is obtained, where V_T is in m s^{-1} and d is in mm, whereas substitution in equation (4.4) yields

$$V_T = 4660\ \mathrm{m}^{2/3} \tag{4.6}$$

where V_T is in m s^{-1} and m is in grams.

4.2 PLUMED SEEDS AND FRUITS

The mode of operation of plumed seeds and fruits is in many respects similar to that of a parachute. In the course of time, plumes have evolved to a configuration which presents a large drag-producing surface for a given volume of biomaterial. The main component of the weight of a plumed disseminule is that of the seed or fruit itself. The disposition of the seed relative to the plumes ensures that, in still air, the fruit falls in a stable configuration with the seed leading and the plumed surfaces advantageously orientated to optimize their drag. As a result of the high ratio of surface area to weight, this group of seeds and fruits has low terminal velocities, typically in the range from 20 cm s^{-1} up to about 60 cm s^{-1}. These airborne bodies are much larger than the tiny particles discussed in the previous section, and so the Reynolds numbers at which they operate are correspondingly higher.

Drag is the aerodynamic force which dominates the motion of this group of disseminules but inertial forces, rather than viscous forces, are the predominant influence.

Because of their low terminal velocity, the plumed seeds and fruits are readily transported horizontally by a slight breeze, and the close proximity of the point of liberation to the ground is not a severe impediment to the achievement of a good dispersal range provided a small upward convection velocity is available. The dandelion (*Taraxacum officinale*) is perhaps the most familiar of all the plants which use this mechanism for dispersal. The plant can sense the ambient humidity, temperature, and wind, and only when conditions are ideal will the fruit finally be released. A warm, dry day with a steady, reliable breeze provides the conditions for which the dandelion waits, and these are most frequently encountered in the early afternoon.

Examples of plants which produce plumed seeds are species of willow-herb (*Epilobium*) and milkweed (*Asclepias*). These seeds develop in a capsular fruit which splits to release the seeds at intervals.

4.3 PLAIN WINGED SEEDS

There is a group of seeds, chiefly to be found on shrubs, trees, and certain climbing plants, having seed coats which take the form of two thin membranes and between which the embryo is contained. The thin seed coat takes on the role of an aerodynamic lifting surface or wing. Provided the membrane grows perfectly symmetrically the seed is capable of linear gliding flight over a substantial range in still air. Frequently, though, small imperfections arise as a result of uneven ripening and therefore a more common occurrence is for these seeds to follow an irregular flight path from the moment of liberation until they descend to their final resting place.

The analysis of the steady, regular gliding flight of perfectly shaped examples of these seeds is identical to that of aircraft in unpowered flight, the trajectory that the seed follows being determined by the equilibrium of the three forces acting on the seed: its weight, together with the aerodynamic lift and drag (Fig. 4.3). Dispersal of this group of seeds rests on a mechanism quite distinct from the first two groups, previously discussed. The point of liberation is some distance above the ground and, for regular linear flight in still air, the range achieved is directly proportional to the height of the point of liberation above the ground, increasing with the lift/drag ratio of the seed. Seeds which do not manage to disperse outside the 'shadow' of the parent tree—either because they have a poor lift/drag ratio or because they develop a spiralling or zig-zag flight due to asymmetries in geometry—inevitably find it more difficult to germinate and prosper than do seeds that succeed in achieving good range. There is therefore a constant evolutionary pressure in favour of seeds with efficient lifting characteristics which take them beyond the influence of their parents.

Of all the seeds and fruits whose motion depends upon the aerodynamic means of dispersal perhaps the one with the most celebrated qualities is the large seed produced by the exotic Javanese climber, *Zanonia macrocarpa*. The membranes of the wing are shiny and transparent, and typically it has a span of about 15 cm and a mean chord of some 5 cm.

It is instructive to compare the aerodynamics of the *Zanonia* with that of a flat plate, rectangular in plan. For the latter the centre of pressure, that is the position through which the resultant aerodynamic force acts, is towards the trailing edge at low angles of incidence and moves rapidly forward towards the quarter chord position with increasing incidence angle (Houghton and Brock, 1960). There is a corresponding variation in the lift/drag ratio as incidence increases. The maximum lift/drag ratio corresponds to a centre of pressure position just aft of the quarter chord point. To satisfy the requirements of static equilibrium during gliding the centres of mass and of pressure must be coincident.

For a plate made of homogeneous material the centre of mass is at the semi-chord position, and with the centre of pressure in this position a poor lift/drag ratio and a steep glide path are indicated. Furthermore this flight condition is highly unstable. An ordinary rectangular sheet of paper used to represent such a wing readily displays these unsatisfactory characteristics.

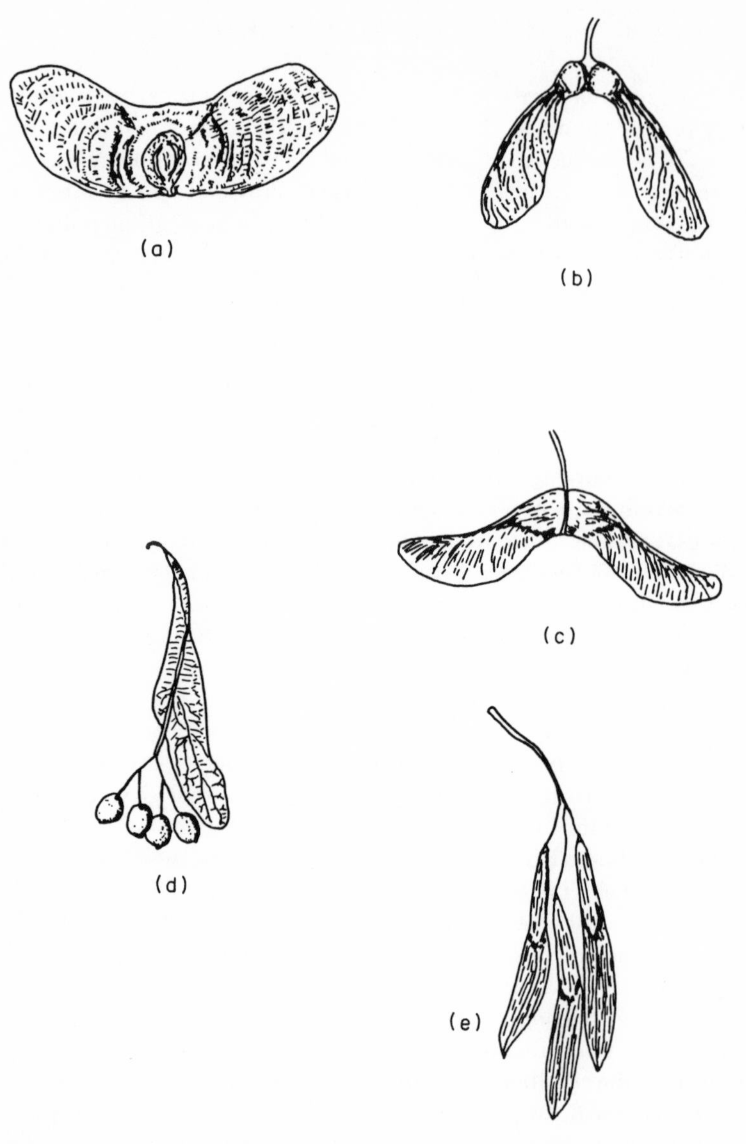

Figure 4.2. Airborne dispersal by means of aerodynamic lift (drawings not to the same scale): (a) the flat seed of the *Macrozanonia* uses gliding flight; (b)–(e) four autorotating samaras: (b) sycamore (*Acer pseudoplatanus*); (c) Norway maple (*Acer platanoides*); (d) lime (*Tilia*); (e) ash (*Fraxinus*).

However, the gliding performance of this thin wing can be substantially improved if its centre of mass is adjusted to a position close to the quarter chord line by, for example, attaching paper clips to its leading edge.

The precise position of the centre of mass determines the angle of incidence at

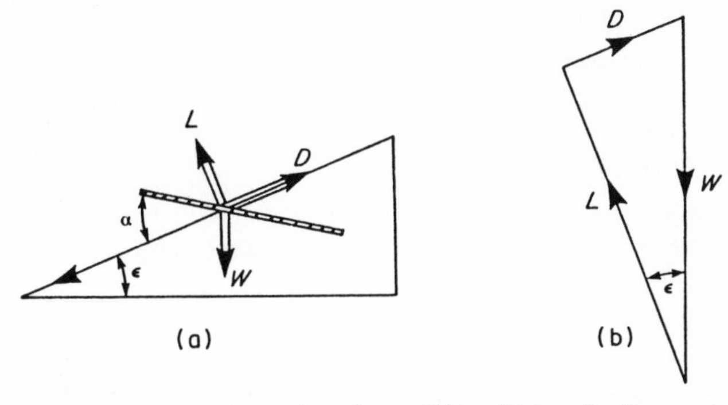

Figure 4.3. Aerodynamics of the gliding flight of a flat seed, such as *Zanonia macrocarpa*: (a) glide path trajectory; (b) triangle of forces.

which static equilibrium is achieved, and in turn this affects the glide angle of the wing, since the tangent of the glide angle is equal to the inverse of the lift/drag ratio (Fig. 4.3(b)). Thus, for a given height of release, the longest range is achieved by having the centre of mass at the position where the centre of pressure corresponds with the maximum lift/drag ratio.

We can now see how these considerations have influenced the evolution of the *Zanonia*'s wing geometry. The largest concentration of mass in the *Zanonia* is the embryo, and this is positioned on the plane of symmetry of the wing and as far forward as is possible. Even so, due to the size of the embryo, the centre of mass is perhaps some 35 per cent aft of the leading edge of the plane of symmetry. For a rectangular wing this would necessitate a rather low angle of incidence for equilibrium, a poor lift/drag ratio and, as a result, the glide path would be unduly steep and unstable. The *Zanonia* has successfully circumvented these undesirable effects by means of sweepback. On each section of the wing the centre of pressure is required to be just aft of the quarter chord point, relative to the local leading edge for efficient gliding. Hence, when the overall aerodynamic load on the wing is taken into account it is evident that the centre of pressure will act through the plane of symmetry, but it will be considerably further aft of the leading edge than it would be on a simple rectangular wing. Thus the sweepback of the *Zanonia* wing allows it to exploit the benefits of operating at or close to an optimum angle of incidence consistent with a small glide angle. Perhaps more importantly, as a consequence of sweepback, the *Zanonia* is blessed with a reasonably stable flight configuration. All of these factors serve to provide the *Zanonia* with a good dispersal range.

4.4 WINGED FRUITS AND SEEDS WHICH AUTOROTATE

Surely one of the most fascinating evolutionary developments in the plant world is the winged seed or fruit which autorotates as it descends from its point of liberation. As a result of the autorotation the terminal velocity of the samara* is

* The term samara was originally used to describe any winged fruit, but there is an increasing tendency to restrict the term to those fruits which autorotate.

very much less than it would otherwise be. This evolutionary process has occurred amongst the fruits of a range of trees, including the lime, maple, sycamore, and ash. This mode of flight has also been developed by the seeds of spruces, pines, and other conifers.

A number of distinct evolutionary configurational trends can be discerned amongst samaras. One line is typified by the fruits of the maple (*Acer platanoides*) and the sycamore (*Acer pseudoplatanus*). A quite distinct trend has been followed by the lime (*Tilia*).

Over one hundred years before mechanical flight was achieved the airborne motion of the samara of the sycamore had been the object of close study by the founding father of aeronautics, Sir George Cayley. Cayley referred to this samara, using the terminology of the day, as the sycamore chat. In his notebook, under the date 9 October 1808, there is an entry which reads:

> I was much struck by the beautiful contrivance of the chat of the sycamore tree. It is an oval seed furnished with one thin wing, which one would at first imagine would not impede its fall but only guide the seed downwards, like the feathers upon an arrow. But it is so formed and balanced that it no sooner is blown from the tree than it instantly creates a rotative motion preserving the seed for the centre, and the centrifugal force of the wings keeps it nearly horizontal, meeting the air in a very small angle like the bird's wing, and by this means the seed is supported till a moderate wind will carry it in a path not falling more than one in 6 from a horizontal one, so that from a moderately high tree it may fly 60 yards before it reaches the earth. The following is the exact figure [there is a sketch in the notebook which is not reproduced here].
>
> In still air it only fell thro' 17 feet in 4 seconds, so that from a tree 30 feet high with a breeze of only 20 miles per hour, a chat would reach the earth at 70 yards from the parent tree.

Cayley's notes indicate that he had a superb eye for observational detail.

The motion of the samaras of the maple and sycamore is in many respects remarkably similar to that of helicopters and autogyros. Indeed it is possible to use helicopter theory (Norberg, 1973) to analyse in considerable detail the motion of such samaras. Since this is rather complicated, we shall content ourselves by summarizing some of the salient details referring to Fig 4.4 on page 62.

In stable vertical descent at constant sinking speed, the total aerodynamic force, F, on the samara is just equal to its weight, W. Viewed from above, the wing of a samara sweeps out a disc of area A_D.

It is advantageous for a samara to sweep as large a disc as it can. This factor has been the evolutionary pressure that has led to the samaras of the maple and sycamore concentrating the mass of the seed at one end of the disseminule since, on purely mechanical grounds, the centre of rotation must be close to, if not coincident with, the centre of mass. Thus in the samaras of the maple and sycamore the centre of mass is no more than 10–20 per cent of the total span from one end.

Experience has shown that in descent through still air samaras may follow one of two different types of trajectory. In the mode of flight commonly found, the centre of rotation is a straight line and is approximately coincident with the centre of mass, so that the wing tip of the samara describes a single helix.

Besides this flight style there is, however, a second, more complex trajectory, during which the samara describes a compound helix, wherein the centre of rotation of the wing itself follows a helical path.

As they descend through still air the samaras of the maple and sycamore do not adopt an essentially horizontal position; rather the seed is at a significantly lower elevation than the tip of the blade. The angle between the mean line through the samara and the horizontal is referred to as the coning angle, β. This angle plays an important role in stabilizing the rotational speed of the samara. Centrifugal forces acting on the rotating wing produce a turning moment which tends to make the wing adopt a horizontal position; aerodynamic forces give rise to a turning moment which tends to move the wing into a vertical position. The coning angle is determined by the equilibrium condition when the moments due to centrifugal and aerodynamic forces are equal and opposite.

We shall now consider briefly the relative air velocities which are experienced by a samara as it descends vertically through still air. From Newton's laws of motion we can deduce that in order to produce the total aerodynamic force, F, in the upward direction, the air through which the samara passes experiences a total rate of change of momentum equal in magnitude to F but in the downward direction. In other words the samara drags some air down with it. As a result the vertical component of the velocity of the air relative to the samara is not simply equal and opposite to the sinking speed, V_s, but has some other value which we shall denote by V_D, which is rather less than V_s. The value of V_D in fact varies over the span of the samara. The relative vertical velocity V_D can itself be resolved into two components: the first along the span of the blade, V_{DS} and the second normal to the quarter chord line of the blade, V_{DN}. In practice the component V_{DS} makes no significant contribution to the aerodynamic forces acting on the blade.

Due to the rotational motion of the samara, there is a tangential velocity component V_t which varies across the span and is given by

$$V_t = \omega r \tag{4.7}$$

where ω is the angular velocity of the samara and r is the radial distance of the blade section from the rotation axis.

Consequently the resultant velocity of the air relative to the blade, V_R, is given by

$$V_R = \sqrt{(V_{DN}^2 + V_t\text{u}22)} \tag{4.8}$$

which can be written in the alternative but equivalent form

$$V_R = \sqrt{(V_D^2 \cos^2\beta + \omega^2 r^2)} \tag{4.9}$$

The effective incidence of the blade, α, at any section is the angle between the resultant air velocity vector and the line of the mean chord. Referring to Fig. 4.4, $\alpha = \psi - \theta$, where the angle between the chord line and the horizontal, θ, is approximately constant over the span.

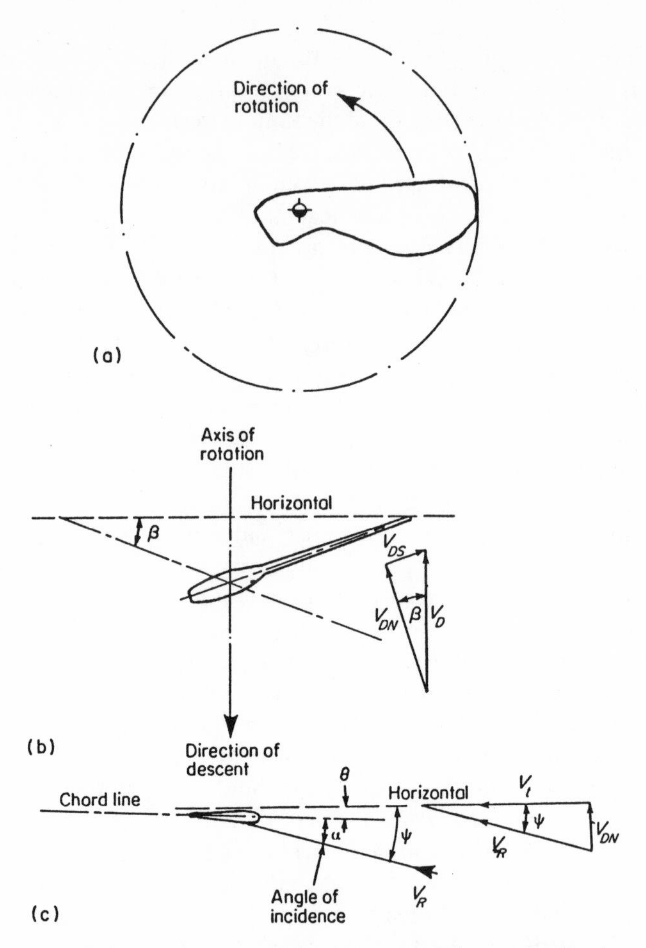

Figure 4.4. The autorotating descent of the maple (*Acer*) samara through still air: (a) plan view; (b) side elevation; (c) view through a section of the samara, looking along the span.

From Fig. 4.4

$$\tan \psi = \frac{V_{DN}}{V_t} = \frac{V_D \cos \beta}{\omega r} \tag{4.10}$$

Equation (4.9) shows that the relative air velocity V_R increases substantially with r, and equation (4.10) shows that the angle ψ decreases with increasing r. Both of these factors indicate that the aerodynamically important parts of the blade are towards the blade tip. At the extreme tip, due to so-called tip losses, the blade cannot support any aerodynamic load. Just inboard, however, the angle of incidence, α, is modest so that the sections in this region can generate high lift/drag ratios and due both to the high velocity of the air relative to the blade, and to the size of the chord, the main contributions to the overall aerodynamic force come from these parts of the blade. Close to the axis of rotation, where r is small, the tangential velocity component is negligible, so that the velocity of the air relative to the blade is essentially in the vertical

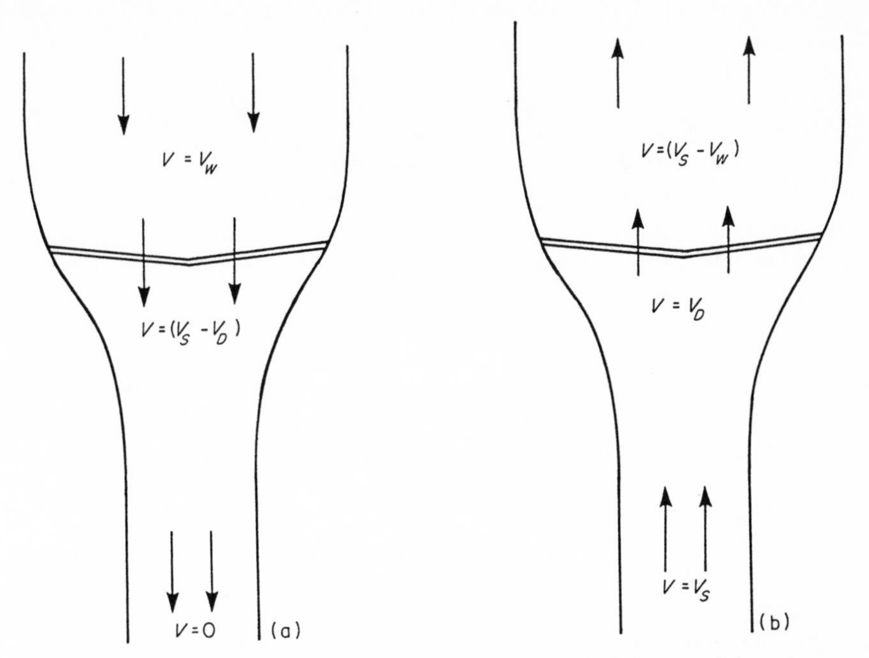

Figure 4.5. Schematic representation of samara in vertical descent: (a) stationary coordinate system; (b) coordinate system moving with samara.

direction. Thus the contribution of this part of the blade to the overall aerodynamic force is in the form of drag, since even if the blade were suitably profiled in this region, no worthwhile vertical component of lift would be generated. Hence the lack of any aerodynamic optimization by nature in this region is fully explained. In contrast, where optimization is required, so it exists: the variation of the chord length across the outer part of the span shows a remarkable similarity to the loading distribution on the blades of windmills and propellors.

The preceding discussion was concerned with the local variations of the aerodynamic properties across the span of the samara. Some useful insights into the samara's aerodynamic qualities can be obtained by a different approach which focuses attention on its overall behaviour. Figs 4.5(a) and 4.5(b) are schematic representations of the samara falling through still air. It is assumed that the air influenced by the samara can be represented by the momentum jet shown. In Fig. 4.5(a) the velocities in the jet are described relative to a stationary frame of reference. In Fig. 4.5(b) the same situation is illustrated, but here the velocities are indicated with respect to a frame of reference descending at the sinking speed of the samara. To analyse the motion of the falling samara we shall apply a technique which has wide application in the study of fluid dynamics, namely actuator disc theory. In the present application of the theory it is assumed that the actuator disc, in this case the autorotating samara, brings about a discontinuous change of pressure from p_1 to p_2 across the disc surface, although no discontinuity in the velocity is assumed across this surface.

In accordance with the usual assumptions made when applying this theory we shall assume:-

(1) swirl and transverse velocity components are negligible;
(2) the velocity across any section of the momentum jet is uniform;
(3) the air well below the samara is at ambient pressure, p_a, and that in the wake well above the samara the pressure reverts to this value.

Referring to Fig. 4.5(b) and applying Bernoulli's equation upstream of the disc

$$p_a + \frac{1}{2}\rho V_s^2 = p_1 + \frac{1}{2}\rho V_D^2 \tag{4.11}$$

where V_D is the velocity of the air relative to the samara at the actuator disc surface, and V_s is the sinking speed of the samara.

Similary, applying Bernoulli's equation downstream of the disc

$$p_a + \frac{1}{2}\rho(V_s - V_w)^2 = p_2 + \frac{1}{2}\rho V_D^2 \tag{4.12}$$

where V_w is the wake velocity in still air, measured in a stationary frame of reference.

Subtraction of equation (4.12) from equation (4.11) yields

$$p_1 - p_2 = \rho V_w\left(V_s - \frac{V_w}{2}\right) \tag{4.13}$$

A force balance on the samara yields

$$W = mg = F = (p_1 - p_2)A_D \tag{4.14}$$

where A_D is the disc area, F is the overall aerodynamic force, and W is the weight of the samara.

Combining equations (4.13) and (4.14) we obtain the important relation

$$\frac{W}{\rho A_D} = V_w\left(V_s - \frac{V_w}{2}\right) \tag{4.15}$$

Momentum considerations require that the weight of the samara equals the overall rate of change of momentum of the air passing through the actuator disc. Thus

$$W = F = \rho Q V_w \tag{4.16}$$

where Q is the volume of air passing through the actuator disc in unit time. Hence

$$Q = V_D A_D \tag{4.17}$$

By combining equations (4.16) and (4.17) to eliminate Q and comparing the resulting equation with equation (4.15) it is established that

$$V_D = V_s - \frac{V_w}{2} \tag{4.18}$$

We assume V_w is linearly related to V_s, so that

$$V_w = KV_s \tag{4.19}$$

where K is a constant of proportionality. Elimination of V_w between equations

(4.15) and (4.19) yields the expression

$$V_s = \sqrt{\frac{2}{K(2 - K)}} \left(\frac{W}{\rho A_D}\right)^{1/2} \tag{4.20}$$

Equation (4.20) indicates that the sinking speed varies as the square root of the disc loading, W/A_D. The importance of a large swept area to the attainment of a low sinking speed is demonstrated.

The magnitude of V_s depends upon the numerical value of K. The sinking speed is a minimum for $K = 1$ and is given by

$$V_{s\min} = \sqrt{2} \left(\frac{W}{\rho A_D}\right)^{1/2} \tag{4.21}$$

For values of K in the range $0.5 \leqslant K \leqslant 1$, the sinking speed given by equation (4.20) departs by no more than about 15 per cent from $V_{s\min}$. Hence equation (4.21) can be expected to provide a useful approximation to the sinking speed of samaras.

In Table 4.2 some results of various measurements on samaras are reported. Included in the table are values of $V_{s\min}$ calculated using equation (4.21), and based upon the measurements of A_D reported by Norberg (1973). It is seen that the values calculated from equation (4.21) are lower than the measured data, as was to be expected, but on this evidence equation (4.21) is shown to be a reasonable basis for a first-order estimation of sinking speed.

Table 4.2 Representative data for some samaras. Predictions based on equation (4.21) are given in the final column. All other values are based on measurements.

Samara	Span (cm)	Mass (g)	Sinking speed ($cm\ s^{-1}$)	Angular velocity ($rev\ s^{-1}$)	Theory, equation (4.21) $V_{s\min}$ ($cm\ s^{-1}$)
Seed of spruce (*Picea abies*)	1.4	0.0075	64	20	60
Fruit of Norway maple (*Acer platanoides*)	4.7	0.13	100	13	74
Fruit of lime (*Tilia*)	—	0.21	230	33	—

The unsteady motion which precedes the entry of a samara into autorotation is rather complicated (Norberg, 1973), and so we shall not attempt to explain it in detail. It is perhaps worth recording, however, that measurements have shown that, prior to autorotation, a maple samara accelerates to a sinking speed some six times the final sinking speed attained under conditions of steady autorotational descent.

The slow sinking speed of maple and sycamores allows them to respond effectively to any horizontal component of wind, which is vital to efficient dispersal.

The flight style of the fruits of the maple and sycamore is shared by the seeds of pines and spruces. These conifer seeds are broadly similar in shape to the samaras of the maple and sycamore but are much smaller and more delicate, probably because they operate in the more sheltered conditions to be found in the densely wooded habitat generally favoured by conifers.

Ash (*Fraxinus*) and tulip (*Liriodendron*) samaras have evolved a flight style even more complicated than that of the maple and sycamore. Like the sycamore samara, ash and tulip keys descend through still air with the centre of mass approximately coincident with a vertical axis of rotation, about which the wing tip moves along a helical path; the long axis of the samara forms a coning angle relative to the horizontal. Careful observation has shown that the keys of the ash and tulip trees rotate about the long axis of the samara in such a way that the underside of the blade advances through the air more rapidly than does the topside (McCutchen, 1977). It is found that most ash and tulip samaras will rotate in either hand, and measurements show that, typically, they spin between three and six times about the long axis for every complete rotation about the vertical axis. The aerodynamic significance of the additional mode of rotation and its consequences for dispersal are incompletely understood and are worthy subjects for further investigation.

The fruit of the lime (*Tilia*) is a superb example of evolutionary development. It has adapted for dispersal by attaching itself to a specialized leaf, known as a bract. Indeed it is usual to find a single bract supporting several seeds. The flight mechanics of the lime samara is similar in principle to that of the sycamore or maple. In still air the disseminule descends through a vertical path with the aerodynamic lift generated on the surface of the autorotating bract which supports its heavy cargo of seeds slung beneath. The shape of the bract really is quite complicated; there is a substantial curvature of the mean chord line so that, for the orientation adopted by the samara in flight, a projection of the wing onto horizontal and vertical planes gives surface areas of comparable magnitude. The point of suspension of the seeds is towards the centre of area of the bract and, as the axis of rotation is again near the centre of mass, the diameter of the disc area swept by the bract is only slightly larger than the projection of the span of the bract onto a horizontal plane. In this sense the aerodynamic efficiency of the lime samara is lower than that of the maple and sycamore samaras, and this is manifested in the lime having a significantly greater sinking speed. But for efficient dispersal samaras do not operate in still air; they must have a breeze. It is in exploiting a horizontal wind that the lime is so efficient. Under such conditions a substantial component of aerodynamic drag can be generated in the horizontal direction, the orientation of the samara being tilted slightly so that the overall aerodynamic force just balances the weight, a necessary condition for steady motion, and the samara descends at a substantial angle to the vertical.

The number of seeds it supports substantially influences the weight and hence the aerodynamic characteristics of the lime samara. Typically the mass is between 0.1 and 0.3 g, sinking velocities range from 100 to 300 cm s^{-1}, and spin rates are from 5 to 40 revolutions per second.

4.5 TUMBLEWEEDS

Before concluding this section, we should not overlook a final class of wind dispersal mechanism adopted by a variety of plants in many parts of the world. Tumbleweeds occur mainly in open land such as deserts, prairies, or steppes, and they rely for dispersal upon the detachment from the parent plant of large pieces which carry the seeds. These are blown along the ground before the wind, rolling over and over as they go, thereby scattering the fruit or seeds. A similar dispersal mechanism is used by some plants whose habitat is the sand dunes near the sea shore.

4.6 AN OVERALL VIEW

From the discussions in the preceding sections the reader will have gathered that the flight mechanisms adopted by different seeds and fruits are very much a function of their size and weight. Pollen, spores, and tiny dust seeds occupy sizes ranging typically between 0.005 mm and 0.1 mm, and all of this group rely on aerodynamic drag forces to gain movement from the parent plant. Their sizes and terminal velocities are such that these small particles move about at Reynolds numbers where their motion is dominated by the viscous properties of the air; the air density has no effect.

Seeds and fruits which are similar to a parachute in flight style are larger in size, typical dimensions ranging from about 0.5 cm to 2 cm, but again aerodynamic drag is used to harness the wind. These bodies travel at rather higher Reynolds numbers, perhaps in the range 10 to 1000. Due to their body shape the major component of drag is normal pressure drag.

The remaining categories of seed and fruit contain the largest and heaviest forms to exploit aerodynamic forces and, in contrast to the previous groups, aerodynamic lift assumes the most important role. There are two separate groups involving different flight styles: the first group contains disseminules which can perform a linear gliding flight; the second contains the autorotating samaras. It is interesting to ponder upon the emergence of the two distinctive styles exploiting aerodynamic lift. The style based on linear gliding flight is very susceptible to small deviations from the perfect conditions required for maximum range. Small asymmetries in the morphology of the disseminule itself, or the occurrence of an isolated small gust during flight, are sufficient to have a dramatic effect on the range achieved. By way of contrast, the autorotating samaras are perhaps less efficient in an aerodynamic sense but they are at the same time less at the mercy of the inherent turbulence in ordinary winds. These considerations are certainly consistent with the different habitat occupied by the plants with the different types of disseminule. Those with plain winged fruits or seeds tend to be restricted to the sheltered conditions in the interior of tropical rain forests, whereas the more open country is favoured by most trees with large autorotating samaras.

Chapter 5

Bird Flight: I—Gliding and Soaring Flight

There are so many remarkable facets to nature that it is perhaps unrealistic to attempt to single out any particular aspect as the supreme manifestation of evolutionary development. Yet of all the fascinating topics which can lay claim to special mention, surely the evolution of bird flight is amongst the most outstanding. Birds are everywhere around us and it is so easy to take the attainment of powered flight for granted. But we can gain some insight into just what an achievement this has been by considering man's own efforts to imitate birds. All attempts by man to fly by securing artificial flapping wings to his own limbs have met with, and indeed were doomed to meet with, failure.

Two main theses have been proposed to explain the evolutionary development of flight. The first explanation suggests that flight evolved progressively in long-tailed, bipedal reptiles which, whilst running rapidly across the ground, derived aerodynamic lift on their fore-limbs. An alternative hypothesis suggests that the first birds were arboreal in habit—having evolved from tree-climbing reptiles in the Jurassic period about 150 million years ago—and used their fore-limbs for the purposes of gliding flight.

Evolutionary advancement seems to have led to the scales on the forelimbs becoming elongated and, in the course of time, they developed into feathers. Over the same period the passive use of the forelimbs was superseded by their utilization in a flapping mode. So little is known about the detailed processes of evolution in those remote times that few words can be set down with confidence. Indeed, it is conceivable that the evolutionary pressures which originally led to the growth of the scales from which feathers then developed was a need for adequate body cooling, and the aerodynamic benefits which were thereby conferred were just an unexpected bonus.

Whatever the impetus for change, a stage was reached where the aerodynamic qualities increasingly presented decisive advantages—in the form of freedom of movement, accessibility of new food supplies, a ready means of escape from predators—and so these aspects became the focus of certain lines of evolutionary change. The skeletal arrangement became progressively modified from its original reptilian form, so that the fore-limbs became entirely responsible for aerial locomotion, the hind-limbs catered for perching, standing, or walking, and the backbone evolved to a form which was able to accept these two distinct modes of supporting the body weight. The characteristic shape and the related life style which we now associate with birds had emerged.

The oldest fossil remains so far discovered, which display the essential characteristics of the birds, are those of Archaeopteryx, five specimens of which have been recovered from rocks of the Jurassic period. In particular Archaeopteryx had feathers. It also had a skeleton which in a broad sense is similar to that of modern birds, but it possessed a long reptilian tail and claws on its wings.

5.1 ANATOMY AND PLUMAGE

In steady, horizontal flight, a bird has to generate sufficient lift to balance its own weight and must also develop a propulsive thrust which is sufficient to overcome the various resistive forces which accompany the bird's motion through the air. These two factors have important consequences upon the anatomy and plumage of birds.

The skeletal structure of birds has evolved to provide strength with a minimum weight penalty. This has been achieved in various ways. Fusion of the bones has occurred, limiting the number of joints at which movement can take place, with a consequent simplification of the muscular system.

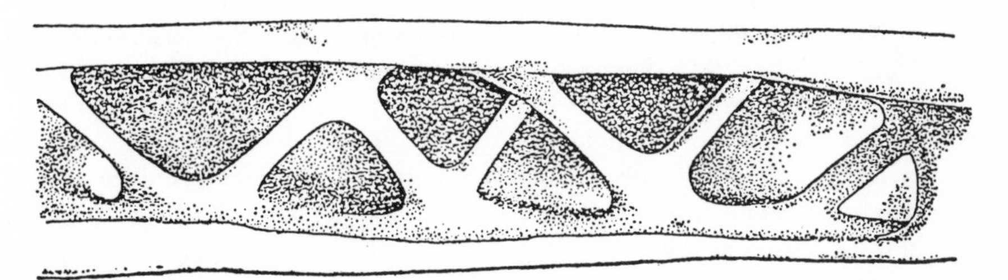

Figure 5.1. Section of the metacarpal bone from a vulture's wing, showing the hollow interior. The engineering structure known as the Warren truss employs a similar arrangement of struts. (After Prochnow and D'Arcy Thompson.)

Numerous bones are hollow and the spaces created are used to augment the air-sacs which are a vital feature of the respiratory system of birds. In the human lung there is an appreciable dead space of unrespired air during each cycle of breathing. In birds this undesirable feature is avoided by the action of the air-sacs. During inspiration, air passes through the lungs and into the air-sacs; this same air is swept through the lungs during the expiration phase of each cycle.

Throughout the animal world heart rate, H, increases with decrease in body mass, m, the relationship being broadly described by the law $H \propto m^{-1/4}$. Consistent with this trend, birds have significantly higher heart rates than are found in man, ranging from about 100 beats per minute in the turkey family (Meleagrididae) to over 600 beats per minute for some species of humming bird (Trochilidae), in comparison with an average figure for man of 72 beats per minute.

The pectoralis major is the muscle responsible for the downstroke of a bird's wing in flapping flight when it is subject to the largest aerodynamic forces. This huge muscle is firmly anchored to the sternum, or breast bone, and is attached to the wing at the humerus. A much smaller muscle, the supracoracoideus, which is also attached to the sternum, elevates the wing during the more lightly loaded upstroke.

The skeleton of a bird's wing is not unlike that of the human arm and hand, the major differences occurring in the latter. Most of the small wrist bones have become fused to give increased rigidity, but a degree of rotational freedom is retained there and at the elbow. The thumb is the support for the alula, or bastard wing, a small group of feathers utilized by some birds to control flight under low speed–high lift conditions. Whereas the second and third fingers have evolved to carry the primary flight feathers, the fourth and fifth digits have disappeared entirely. The secondary flight feathers are attached to the forearm.

Feathers cover almost the entire body of birds. They serve two principal roles: temperature regulation and the provision of flight capability, although they have other functions, such as are exploited in display. The external feathers, in contact with the air when the bird is flying, determine its general shape and are, therefore, known as contour feathers. They affect the bird's resistance to motion, but so far as aerodynamic qualities are concerned, the most important feathers are the primary and secondary flight feathers carried by the wing, which sustain the bird in the air, and the tail feathers, which play an important function in controlling the bird's attitude in flight.

All birds capable of flight have 9, 10, 11, or 12 primary flight feathers. Primaries are referred to individually by number and the count is always made from the inner primary outwards, because it is the outer feathers which have been lost in the course of evolution.

There is a much greater variation in the number of secondary flight feathers amongst the different species of birds and the variations reflect the different flight styles used. Hummingbirds (Trochilidae) have only 6 or 7 secondaries and this is associated with a short forearm and long manus (hand). Most of the passerines (perching birds) have 9 to 11 secondaries, birds of prey have from 14 to 25 and the albatross (Diomedea) has 32. The low numbers of secondaries are found amongst birds relying heavily on flapping modes of flight, in which articulation of the primaries plays a crucial role. Birds which mainly glide or soar depend principally upon classical fixed-wing aerodynamics and the secondary feathers are the source of lift required in such flight styles.

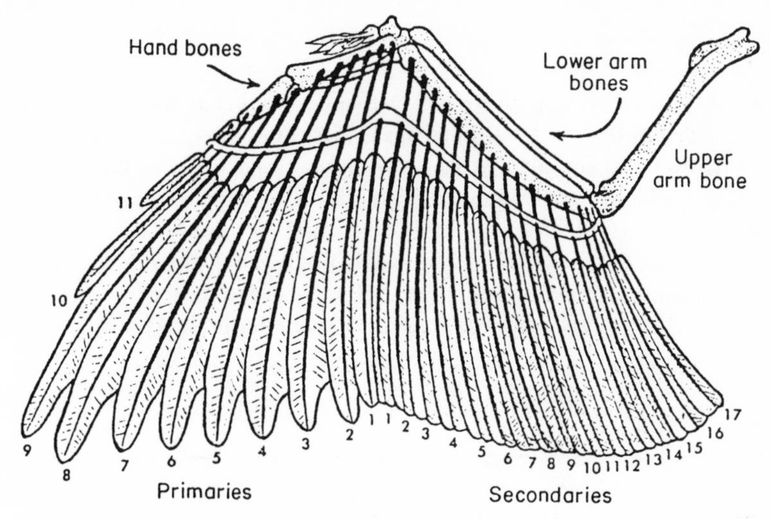

Figure 5.2. The skeletal arrangement and feathers of a typical bird's wing. (Reproduced, with permission of Macmillan Publishing Company, from Hylander, C. J. (1959), *Feathers and Flight*. Copyright © 1959 by Macmillan Publishing Company.)

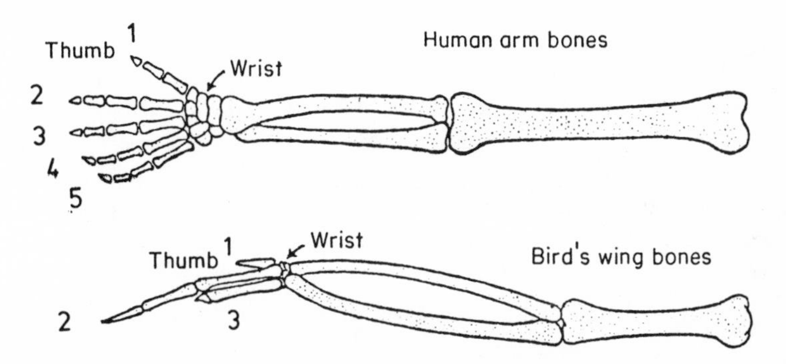

Figure 5.3. A comparison of the bones in a human arm and a bird's wing. (Reproduced, by permission of Macmillan Publishing Company, from Hylander, C. J. (1959), *Feathers and Flight*. Copyright © 1959 by Macmillan Publishing Company.)

5.2 BASIC AERODYNAMICS

There are several distinct modes of flight employed by birds in order to remain airborne: gliding flight, soaring, diving, fast flapping flight, slow flapping flight, hovering. Each of these flight styles involves distinct aerodynamic principles but, before considering them in detail, it is useful to note that they may be classified into two broad groups: flapping flight and non-flapping flight. For the present we shall concentrate on the latter.

Two principal variables largely govern a bird's performance as a flying machine. The first of these is the wing loading, the ratio of the bird's weight, W, to the wing area, S. The second variable is the aspect ratio, A, the ratio of the wing span to the mean chord.

There is a very close parallel between non-flapping flight in birds and the aerodynamics of conventional fixed-wing aircraft, thereby providing a convenient starting point for the discussion of the aerodynamics of bird flight. In a fixed-wing configuration, adopted during gliding or soaring, the total aerodynamic force acting on the bird can be resolved into two components. These are the aerodynamic drag, D, acting along the line of flight and the aerodynamic lift, L, at right angles to the direction of motion. The wing is mainly responsible for the generation of lift, whereas all external surfaces contribute significantly to the drag force.

Table 5.1 Morphological data for a variety of birds.

Bird	Mass (g)	Wing span (cm)	Wing loading ($N\ m^{-2}$)	Aspect ratio
Wren (*Troglodytes troglodytes*)	10	17	24	6.9
Pied flycatcher (*Ficedula hypoleuca*)	12	23	13	5.9
Kestrel (*Falco tinnunculus*)	245	74	35	7.7
Pigeon (*Columba livia*)	330	63	52	6.3
Mallard (*Anas platyrhynchos*)	1100	90	120	9.0
Frigate bird (*Fregata aquila*)	1620	202	50	12.6
Pheasant (*Phasianus colchicus*)	1660	85	104	4.6
Brown pelican (*Pelecanus occidentalis*)	2650	210	58	9.8
Griffon vulture (*Gyps fulvus*)	7300	256	70	6.2
Wandering albatross (*Diomedea exulans*)	8500	340	137	18.7
Andean condor (*Vultur gryphus*)	11700	300	102	7.9

In non-flapping flight a bird derives aerodynamic lift, just as an aeroplane does, by the acceleration of the air over the upper surface of the wing. The high velocities on the upper surface result in the creation of regions of reduced pressure which are primarily responsible for the generation of lift, although the pressure distribution on the lower surface also makes some contribution. The cross-section of the bird's wing varies over the span and the local contribution to the overall lift depends in a complex way on the chord length, the thickness and camber distributions of the wing section, and on the local angle of incidence.

Viewing the wing as a whole, the magnitude of the lift force depends on the average angle of incidence of the wing relative to the airstream, on the relative velocity between the wing and the airstream, V, and on the shape and size of the wing.

As has been shown in the Introduction, the lift, L, of a wing is given by the relation

$$L = \frac{1}{2} \rho V^2 S C_L \tag{5.1}$$

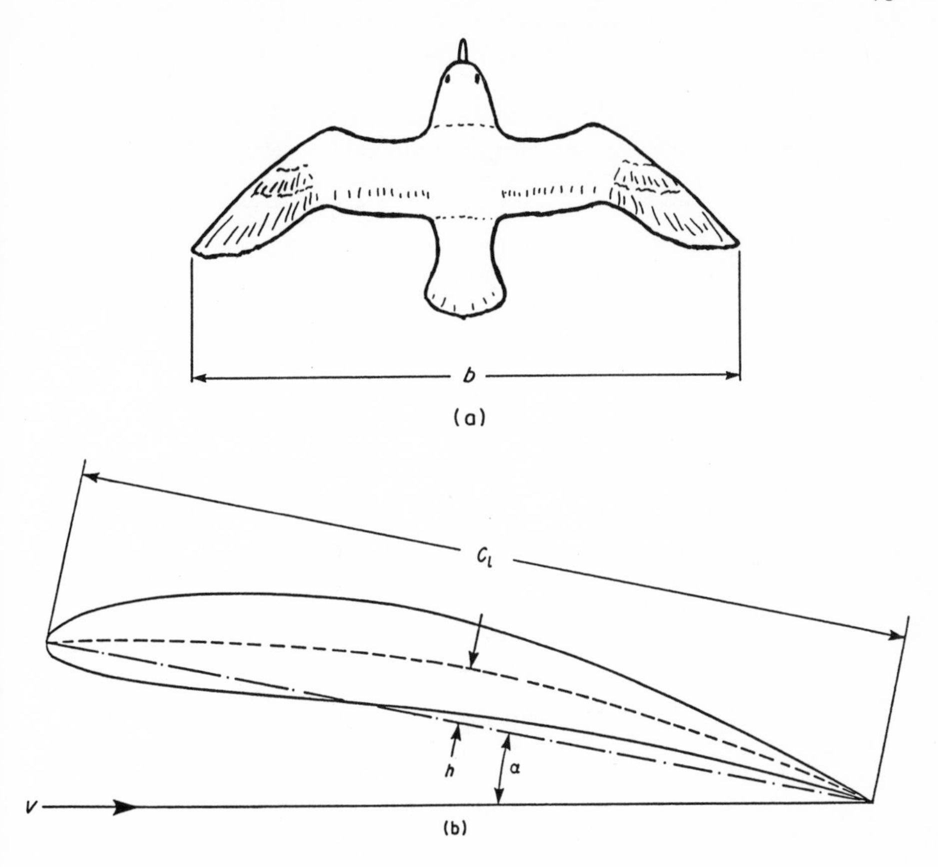

Figure 5.4. Basic geometry of a bird's wing: (a) wing geometry; (b) sectional geometry: α = incidence angle; c_l = local chord; b = wing-span; S = gross wing area; c = mean chord, S/b; A = aspect ratio, b/c; h/c_l = camber.

where S is the wing area and ρ is the air density. The lift coefficient, C_L, increases linearly with the mean angle of incidence of the wing up to just below the angle at which stall occurs.

In all forms of bird flight, with the exception of steep dives and briefly during violent manouevres, the aerodynamic lift is used principally to support the bird's weight. This fact provides a ready method of quantifying, with reasonable accuracy, the lift force sustaining flight.

As with an aircraft, so a bird's resistance to motion, its drag, D, can be expressed in coefficient form by

$$D = \frac{1}{2} \rho V^2 S C_D \tag{5.2}$$

where C_D is the drag coefficient.

The derivation of lift is accompanied by a corresponding penalty in the form of increased air resistance, and so the drag coefficient may be considered in two

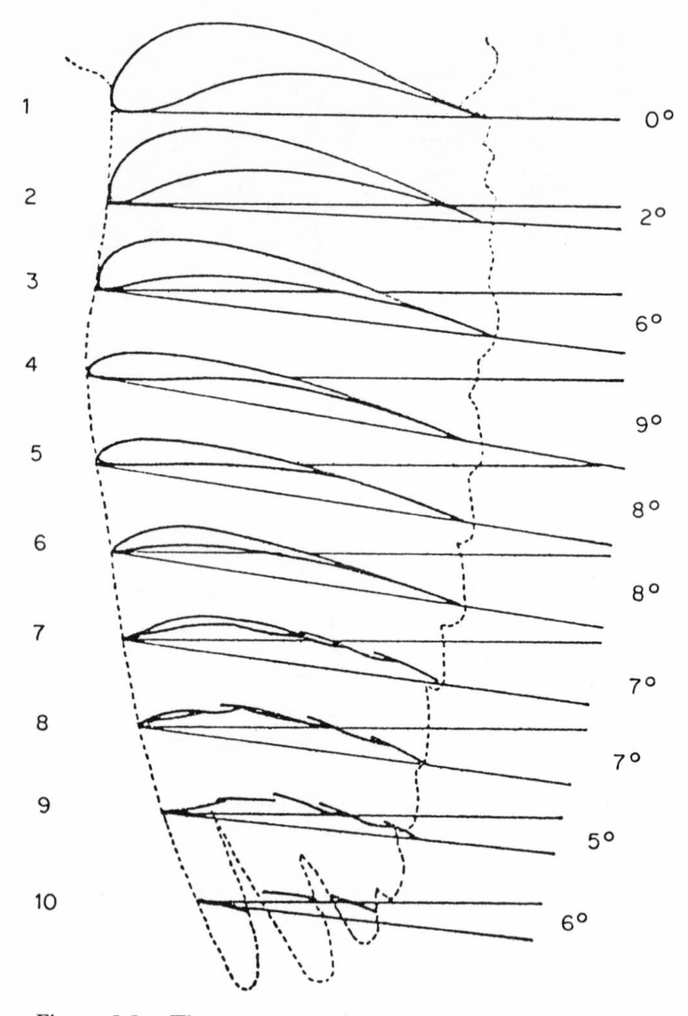

Figure 5.5. The geometry of a pigeon's wing. The cross-sectional shape is shown at different spanwise positions. Note the appreciable camber of the aerofoil sections. Also the wing has a complex twist, with sections towards the centre of the wing having the largest angles of incidence. (Reproduced, with permission from Nachtigall and Wieser (1966), *Z. Vergl. Physiol.*, **52**, 333–346.)

parts: the drag coefficient in the absence of lift, and the increment in drag coefficient consequent upon the generation of lift.

At zero lift the resistance is entirely in the form of profile drag, both the bird's body and wing contributing to this component of drag*. From an aerodynamic

* The term parasite drag was previously used to describe the contribution to profile drag from sources other than the wing. Such terminology is no longer generally employed in an aeronautical context, although it seems to persist in the biological sciences.

viewpoint birds are reasonably streamlined and so the major contribution to profile drag is from skin-friction drag.

When a fixed wing generates lift, a pair of contra-rotating vortices are formed with centres roughly coinciding with the wing tips. At the wing tips there is a constant tendency for the air to flow from below the wing into the low-pressure region created above it, and it is as a result of this motion that, as the wing moves through the air, the vortex pair is left trailing behind the wing. The so-called trailing-vortex drag results from the fact that the wing must constantly do work in order to create the rotational kinetic energy in the vortex.

The lift generated by a wing is not evenly distributed across the span. On the contrary, the local contribution to lift must fall to zero at the wing tips, since there can be no discontinuities of pressure in the air adjacent to the wing tips. The precise form of the loading distribution depends upon the planform and sectional geometry of the wing, and the angle of incidence at which it is operated. The shape of the spanwise distribution curve affects the magnitude of the trailing-vortex drag. When the loading curve is in the shape of an ellipse the trailing-vortex drag is minimized. For this reason the elliptical distribution of lift across the span is of some theoretical importance in aerodynamics.

Although the trailing-vortex drag is the major component of lift-dependent drag, there is a further substantial contribution, from the profile drag, which also increases with incidence. This results from the fact that, with increase in lift coefficient, the conditions under which the boundary layer grows on the upper and lower surfaces of the wing become increasingly inhospitable as the trailing edge is approached.

Over the range of incidences where the lift coefficient and incidence angle vary linearly, it is found that the drag coefficient increases as the square of the lift coefficient and is inversely proportional to the aspect ratio of the wing, and so the drag coefficient can be written in the form familiar to aerodynamicists, which is

$$C_D = C_{D0} + \frac{KC_L^2}{\pi A} \qquad (5.3)$$

where C_{D0} is the drag coefficient at zero lift and the second term is the lift-dependent drag coefficient. In this expression A is the aspect ratio, K is a constant which typically has a value of between 1 and 2, and π is equal to 3.142.

A new equation for the drag can be obtained by combining equations (5.1), (5.2) and (5.3) to give

$$D = \frac{1}{2}\rho V^2 S C_{D0} + \frac{KL^2}{\frac{1}{2}\rho V^2 S \pi A} \qquad (5.4)$$

This equation may, on first impressions, look rather forbidding, but it is not too difficult to understand; furthermore it is a most important relation and will repay careful inspection.

Consider the gliding flight of a given bird over a range of speeds. Without any significant loss of accuracy the lift, L, can be replaced by the bird's weight, W. For

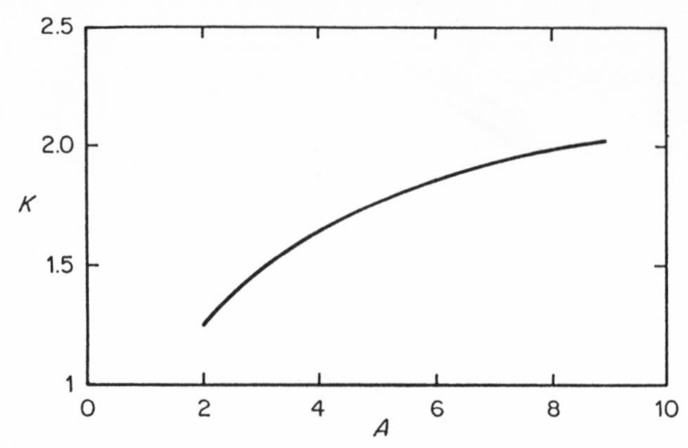

Figure 5.6. Variation of lift-dependent drag factor, K, with aspect ratio. The results are for wings of rectangular plan form, but should provide a useful guide for birds.

a given bird flying with a fixed-wing geometry W, S, A, and C_{D0} are all constant as, of course, are π and ½. The density of the air, ρ, may also be taken as constant. So the only quantity that is varying is the speed, V. It is evident that the first term on the right-hand side of equation (5.4)—the zero-lift drag—increases as V^2, whereas the second term—lift-dependent drag—is inversely proportional to V^2. The zero-lift drag increases with V; the lift-dependent drag decreases with increase in V. As a consequence of these opposite tendencies there is a particular value of V at which the drag is a minimum. The speed at which this occurs is known as the minimum drag speed and is denoted by the symbol V_{md}. At the condition of minimum drag the zero-lift drag and the lift-dependent drag components of equation (5.4) are in fact equal. By equating these two terms an expression for the minimum drag speed is immediately obtained. It is

$$V_{md} = \left(\frac{K}{\pi A C_{D0}}\right)^{1/4}\left(\frac{2L}{\rho S}\right)^{1/2} \tag{5.5}$$

We shall return to the significance of V_{md} subsequently on a number of occasions.

The product of drag and velocity, $D \times V$, is a measure of the power which is required to overcome the resistance of the air. Just as there is a speed for which the drag is a minimum, there is similarly a particular airspeed, denoted by V_{mp}, at which the power is a minimum. Multiplying equation (5.4) by V, the product $D \times V$ is obtained. Using simple calculus it is easily shown that the minimum power speed and the minimum drag speed are related by

$$V_{mp} = \frac{V_{md}}{(3)^{1/4}} = 0.76 V_{md} \tag{5.6}$$

In order to maintain itself in the air a bird must generate an aerodynamic force sufficiently large to support its weight. Otherwise, put simply, it will fall out of the sky.

There is therefore an absolute lower limit to the flight speed, determined by the

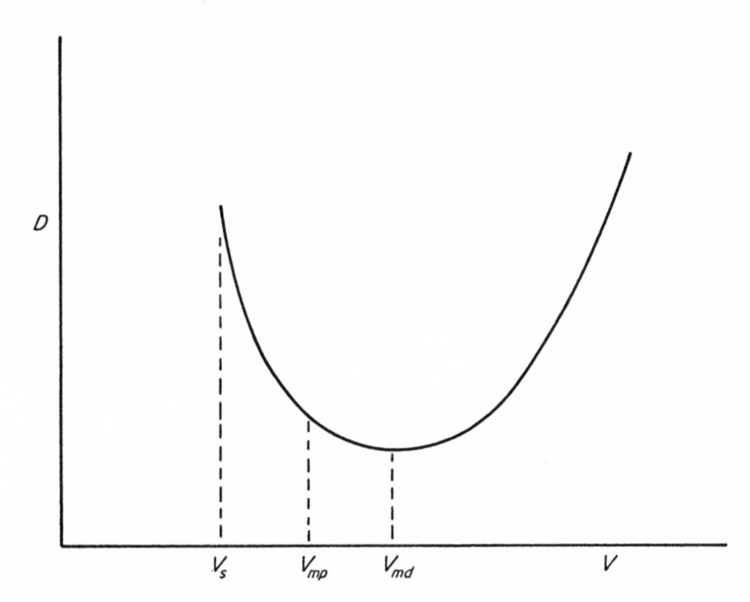

Figure 5.7. Variation of drag with velocity in steady gliding flight. V_s = stalling speed; V_{mp} = minimum power speed; V_{md} = minimum drag speed.

ability of the wings to generate sufficient lift and below which a bird cannot fly without flapping its wings. Denoting the maximum lift coefficient which a fixed wing can sustain in steady flight by $C_{L\max}$, then the minimum airspeed is equal to the stalling speed, V_s, given by

$$L = \frac{1}{2} \rho V_s^2 C_{L\max} S \tag{5.7}$$

or

$$V_s = \left(\frac{2L}{\rho S C_{L\max}} \right)^{1/2} \tag{5.8}$$

In a fixed-wing configuration a representative value for $C_{L\max}$ is about 1.5, although some birds with sophisticated wing geometries can generate values of $C_{L\max}$ as high as 2. At sea level the atmospheric density is about 1.21 kg m^{-3}. Substituting the weight, W, for L, equation (5.8) can be used to calculate the approximate stalling speed of any bird, provided its wing loading, W/S is known. The stalling speed represents the lowest flight speed at which a bird can maintain itself in the air without flapping its wings. To remain airborne below this speed a bird must, perforce, resort to flapping flight. Some representative data are given in Table 5.2.

Substituting W for L in equations (5.5) and (5.8), an important deduction can be made concerning the characteristic flight speeds V_{md}, V_{mp} and V_s. All of these quantities vary as the square root of the wing loading, W/S.

Table 5.2 Approximate values of the stalling speed. At flight speeds above this value the bird can maintain itself in the air without flapping the wings. Below it, flapping is required. Calculations assume $C_{L\max} = 1.5$.

Bird	Stalling speed (m s^{-1})
Wren	5.1
Pied flycatcher	3.8
Kestrel	6.2
Pigeon	7.6
Mallard	11.5
Frigate bird	7.4
Pheasant	10.7
Griffon vulture	8.8
Wandering albatross	12.3
Mute swan	14.8

5.3 GEOMETRICAL SCALING LAWS

It is possible to make some very useful deductions concerning the flight of birds by the application of what are known as scaling laws.

To introduce these ideas let us consider, first of all, a simple example. Let us compare the geometrical properties of two cubes, the side of the larger cube being twice that of the smaller cube. Then the external surface area of the larger cube is four times the surface area of the smaller cube, and the volume it occupies is eight times greater. This result can be generalized. If objects which are geometrically similar in shape are compared then, no matter how complex that shape, the relationship between the surface areas and volumes of those shapes can be computed from a knowledge of the ratio of corresponding linear dimensions of each of the objects. If l is a typical linear dimension, then the surface area, S, varies as l^2 and the volume, $\mathcal{V}$, varies as l^3.

The mass, m, of any object is equal to the product of its volume and density. Then, comparing objects of the same density, it follows that the mass, m, and weight, W, vary with $\mathcal{V}$ or, using the earlier result, with l^3.

Applying these ideas to geometrically similar birds, it is deduced that the wing loading, W/S, varies in direct proportion to l. As it has already been demonstrated that the characteristic flight speeds—the minimum power speed, the minimum drag speed, the stalling speed—all vary as the square root of wing loading, it is concluded that the flight speeds of geometrically similar birds vary as $l^{1/2}$.

Finally we can calculate the variation of Reynolds number with characteristic length, l. For a given kinematic viscosity the Reynolds number is determined by the product of velocity and characteristic length. Since flight speed varies as $l^{1/2}$, it follows that the Reynolds number varies as $l^{3/2}$.

5.4 GLIDING FLIGHT

It is convenient to commence the discussion of particular flight styles by looking at gliding, because in this mode of flight the wings are held in a fixed configuration.

Initially we shall assume the air to be still and the bird to have a steady velocity V relative to the air. During gliding flight the total aerodynamic force is equal and opposite to the weight of the bird, W. The total aerodynamic force may be resolved into two components: the aerodynamic drag, D, and the aerodynamic lift, L. By considering the equilibrium of these forces we deduce that the bird can only maintain a constant velocity, V, along the glide path by losing height, at a sinking speed, v_s, given by $V \sin \varepsilon$.

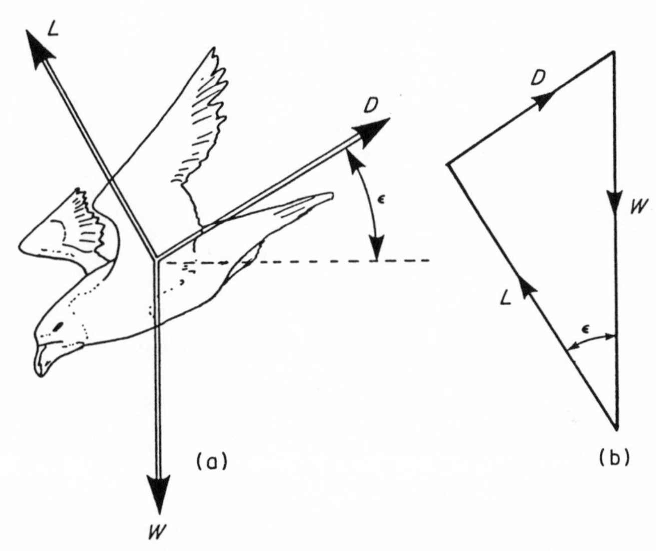

Figure 5.8. Bird in steady gliding or diving flight through still air. Sinking speed $v_s = V \sin \varepsilon$: (a) glide path geometry; (b) triangle of forces.

The angle ε between the glide path and the horizontal is given by

$$\tan \varepsilon = \frac{D}{L}$$

so that in order to minimize the glide path angle a bird must operate at the maximum lift/drag condition. For small glide angles the lift and weight are nearly equal ($L = W \cos \varepsilon$) and since the weight is a constant, it follows to a very good approximation that the minimum glide angle corresponds with the minimum drag condition.

The flight speed appropriate to the minimum glide angle is known as the best glide speed, denoted by V_{bg}. Thus, with negligible loss of accuracy, we can equate the best glide speed to the minimum drag speed, V_{md}, which we have shown is given by

$$V_{bg} = V_{md} = \left(\frac{K}{\pi A C_{D0}}\right)^{1/4}\left(\frac{2W}{\rho S}\right)^{1/2} \tag{5.9}$$

where W has now been substituted for L.

In gliding flight, due to the aerodynamic drag, a bird dissipates energy at a rate given by the product of drag and airspeed. So that there is no reduction in the airspeed, and hence no depletion of the kinetic energy, the bird loses height. There is therefore an associated loss of potential energy at the rate given by the product of the bird's weight and the sinking speed, v_s. The rate of loss of energy can therefore be expressed mathematically as

$$W \times v_s = D \times V \tag{5.10}$$

and this relation holds for all gliding conditions.

In order to minimize the rate of energy expenditure it is advantageous for a bird to glide, not at its minimum angle of descent, which corresponds with the minimum drag speed, but instead at some other speed, thereby minimizing the sinking speed, v_s. The flight speed corresponding to the minimum sinking speed is denoted by V_{ms}. Since W is a constant, it is evident from equation (5.10) and the derivation of equation (5.6) that the velocity V_{ms} is in fact exactly equivalent to the minimum power speed, defined previously. Thus

$$V_{ms} = V_{mp}$$

It is common to represent gliding performance in the form of a graph of v_s against V, and a typical plot is shown in Fig. 5.9.

For velocities V less than $V_{md}(= V_{bg})$ stable gliding flight cannot be achieved with the wings held in a fixed configuration. This can be demonstrated in the following way.

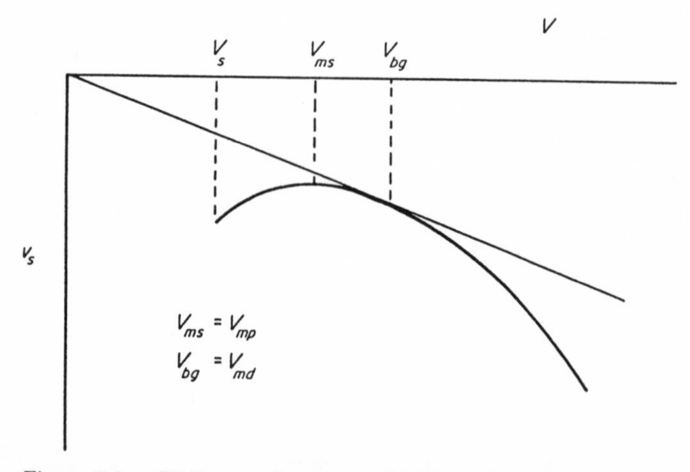

Figure 5.9. Gliding performance of a bird. Sinking speed as a function of airspeed.

Suppose that the bird is gliding with velocity $V < V_{md}$. If now the motion suffers a perturbation reducing the velocity by a small amount δV, then there is a corresponding increase in drag, δD, which tends to reduce the velocity still further. If this process continued indefinitely the stalling speed V_s would be reached. A perturbation increasing the velocity by δV reduces the drag by δD, tending to increase the velocity towards the minimum drag speed. In practice, stable gliding flight with the wings maintained in a rigidly fixed configuration can only be attained at speeds in excess of the minimum drag speed (Lighthill, 1974).

This type of limitation places serious restrictions on the flight of fixed-wing aircraft, but it seems that birds can overcome the inherent stability problem at speeds below the minimum drag speed by drawing on a resource not generally available to aircraft.

The vital distinction rests on the fact that birds' wings are not rigid; their geometry can be changed in two separate respects. Firstly, birds are able to adjust their wing geometry by small amounts using active muscular control. Secondly, their wing feathers deflect appreciably when under load and so, as a consequence of aeroelastic effects, any perturbation of flight conditions automatically brings about some change in the geometry of the wing. It is not known which of these mechanisms birds employ to maintain essentially stable gliding conditions, but there is now substantial evidence that birds glide, in a controlled fashion, at speeds below the minimum drag speed.

This capability is of course of great benefit, since the minimum power speed—the optimum glide speed at which the rate of energy expenditure is minimized—is in the regime where the flight is unstable for a rigid, fixed-wing configuration.

From our earlier theoretical analysis it is evident that for a bird to be an efficient glider in still air it must have a low minimum drag speed, and this condition requires either a high aspect ratio wing geometry or a low wing loading. We are now in a position to compare such theoretical predictions with the state of things in the real world.

There are numerous sea birds which resort to periods of gliding—examples are the gulls (Laridae)—and these birds are characterized by the possession of high aspect ratio wings. Many birds of prey are efficient gliders and, for their size, these birds have large wing areas and, consequently, low wing loading. Thus there is seen to be accord between the aerodynamic parameters of birds which exploit gliding flight and the theoretical requirement of this flight style.

However, gliding is not restricted solely to birds with high aspect ratio wings or lightly loaded wings. In particular just before landing many birds—such as the pigeons (Columbidae), pheasants (Phasianidae), and ducks (Anatidae)—pass through a gliding phase.

We shall conclude our analysis of gliding flight by demonstrating how a bird can change its gliding performance by modifying its wing planform (Fig. 5.10). The wing area, S, is equal to the product of wing span (b) and mean chord (c); aspect ratio, A, is equal to the ratio of wing span to mean chord, b/c. It therefore

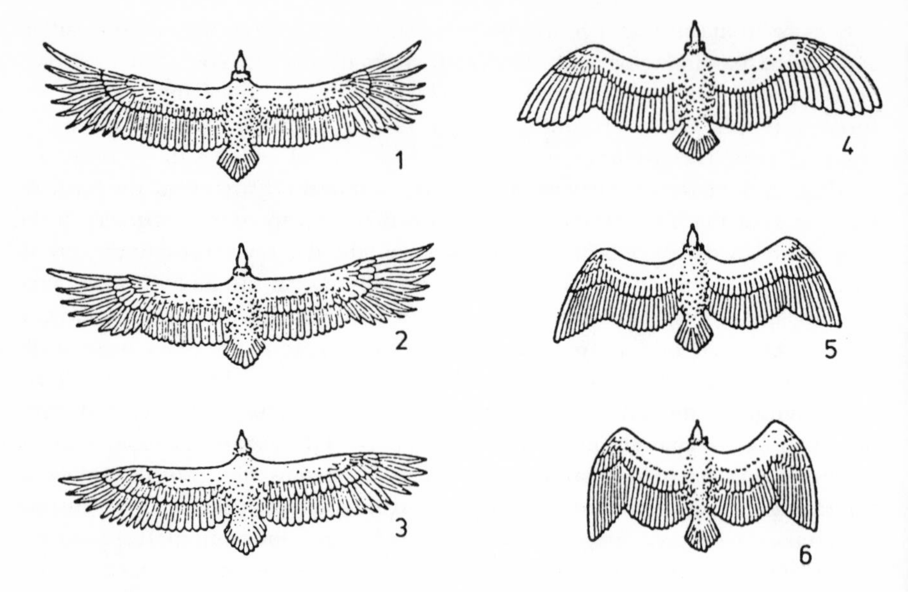

Figure 5.10. Wing configuration of the vulture (*Gyps*) and its variation with airspeed. The highest airspeed corresponds to configuration (6). In order to reduce airspeed the wing span and plan area are increased (5) and (4). Further reduction of airspeed is achieved, (3) and (2), by partial separation of the primary feathers, as well as by increasing still further the basic wing area. At the lowest airspeed (1) the wing is fully extended and the primaries are separated to the maximum extent. (After Ahlborn.)

follows that equation (5.5) can be written in the revised form

$$V_{bg} = V_{md} = \left(\frac{K}{\pi S C_{D0}}\right)^{1/4}\left(\frac{2L}{\rho b}\right)^{1/2} \tag{5.11}$$

In this expression, for small glide angles, L may be replaced by W which is a constant, π is a constant and, at constant altitude, ρ is also constant. Variation of K and C_{D0} are assumed sufficiently small that in the present context they may be neglected. The factors remaining are the wing area, S, and the wing span, b. From equation (5.11) it is deduced that a bird can control its gliding speed primarily by modifying its wing span, which also has a consequential effect on the wing area. An increase in gliding speed, which results in an increase in the related glide angle, is obtained by a reduction in wing span and wing area, and vice-versa. Observations of the flight speeds and wing planforms of various birds, including gulls, pigeons, and vultures, show that this effect is exploited in practice.

During gliding flight in still air, a bird is losing height and hence potential energy. Regaining its original altitude involves a bird in energy expenditure under the more stringent conditions of flapping flight. Large birds in particular find these demands difficult to meet and they have discovered alternative means of gaining or maintaining height by extracting energy from the wind using the methods of slope, thermal, and dynamic soaring.

5.5 DIVING FLIGHT

In gliding flight the lift force largely counterbalances the weight of the bird. As we have seen the angle of glide can be increased by controlled reduction of the wing area, with a consequent reduction in the lift force, and increases in airspeed and in sinking speed result. Under certain conditions some birds take this process virtually to an extreme, almost completely retracting the wings, thereby retaining only a small component of aerodynamic lift. The weight of the bird and the resultant aerodynamic force, which is now almost entirely drag, are no longer in balance and the bird enters a steep dive, plummeting towards the earth and accelerating rapidly. It is possible to estimate an upper limit to the speed that might arise in the course of such a manoeuvre. The extreme case corresponds to a vertical dive, with the bird having reached an equilibrium terminal velocity, at which condition the aerodynamic drag exactly balances the weight. Thus

$$D = \frac{1}{2}\rho V_{max}^2 S C_{D0} = W$$

or

$$V_{max} = \left(\frac{2W}{\rho S C_{D0}}\right)^{1/2} \qquad (5.12)$$

In evaluating this equation the values of reference area, S, and zero-lift drag coefficient appropriate to the bird's configuration with wings retracted, must be used.

Various birds of prey adopt an aerial manoeuvre similar to that just described. Of these perhaps the most celebrated is the peregrine falcon (*Falco peregrinus*), which kills its prey in the air by striking from a steep dive, known as the stoop.

The highest reliably recorded speed for the peregrine in the stoop is 37 m s^{-1} (82 mph). However, there is every reason to believe that the peregrine is capable of reaching much higher speeds during this particular aerial manoeuvre. Careful wind tunnel tests at airspeeds up to 16 m s^{-1} have been made with the laggar falcon (*Falcon jugger*), a native of India, which is similar in size and proportions to the peregrine. These tests (Tucker and Parrott, 1970) involved the investigation of the gliding flight of a live bird over a range of airspeeds and glide angles. Consequently the falcon adopted a range of wing geometries, and for each configuration it was possible to calculate the bird's drag coefficient and reference area. The minimum value of the product SC_{D0} found for any configuration was 3.1×10^{-3} m^2, and substitution of this value in equation (5.12) provides an estimate of the maximum speed a falcon might attain in nature. With $W = 5.6$ N and $\rho = 1.21$ kg m^{-3} a value of $V_{max} =$ 55 m s^{-1} (120 mph) is calculated. In free flight it is possible that the falcon could retract its wings more completely to adopt configurations which could not be accommodated in the wind tunnel tests. By extrapolating the wind tunnel test data a minimum value of $SC_{D0} = 8.9 \times 10^{-4}\ \mathrm{m}^2$ is calculated for the configuration in which the wings are completely retracted, indicating that a terminal velocity as high as $100\ \mathrm{m\ s^{-1}}$ (220 mph) could be attained in the stoop.

5.6 SLOPE SOARING

The discussion of gliding flight considered the air to be a stationary medium through which the bird moved with velocity V and at an angle ε relative to the ground. When an airstream of velocity, V, is inclined upwards at an angle ε to the horizontal, then a bird gliding under these conditions would appear motionless relative to the ground. More generally, a bird will glide relative to the air at a velocity different from that of the airstream relative to the ground, so that the bird will appear to be moving at some intermediate speed relative to the observer on the ground. (Incidentally, this factor makes the interpretation of field studies of gliding and soaring birds notoriously difficult.) Airstreams having the desired property of an upwards component of velocity are frequently to be found on the windward side of cliffs and mountains, and various species of bird have learnt to exploit these conditions to remain airborne, extracting energy from the wind by means of slope soaring. As the discussion of gliding flight has already revealed, birds with low wing loadings and/or high aspect ratio wings, having a low minimum drag speed, are best able to exploit these conditions. Examples are crows, ravens, gulls, and various birds of prey.

Using the preceding analysis for gliding flight it is possible to determine the minimum updraught velocity, v, required for a bird to maintain altitude. The horizontal component of wind velocity is assumed arbitrary, and simply affects the groundspeed of the bird. Initially, flight at an airspeed equal to the minimum drag speed, V_{md}, will be assumed. At this condition the total drag is equal to twice the zero lift drag. Hence

$$D = 2 \times \frac{1}{2} \rho V_{md}^2 SC_{D0} \tag{5.13}$$

and

$$\sin \varepsilon = \frac{D}{W} = \frac{v_{md}}{V_{md}} \tag{5.14}$$

So

$$v_{md} = \rho \frac{V_{md}^3 SC_{D0}}{W} \tag{5.15}$$

Substitution for V_{md} using equation (5.5), with W substituted for L, yields the final result

$$v_{md} = (2)^{3/2}\left(\frac{K}{\pi A}\right)^{3/4}\left(\frac{W}{\rho S}\right)^{1/2} C_{D0}^{1/4} \tag{5.16}$$

Equation (5.16) reveals that for a bird to exploit low updraught velocities any or all of the following factors are required: low wing loading, high aspect ratio, low zero lift drag-coefficient.

The above analysis assumes flight at the minimum drag speed. If a bird is able to maintain controlled flight at an airspeed equal to the minimum power speed, V_{mp}, then analysis shows that it will be capable of sustained horizontal soaring flight at the reduced updraught velocity, v_{mp}, given by

$$v_{mp} = \frac{v_{md}}{1.14} \tag{5.17}$$

A sample calculation reveals the magnitudes involved.

Taking figures which might be representative of a vulture: $A = 6.5$, $W/S = 70\ \mathrm{N\ m^{-2}}$, $K = 1.9$, $C_{D0} = 0.03$ and with a density $\rho = 1.21\ \mathrm{kg\ m^{-3}}$, corresponding to normal sea level conditions, we evaluate

$$v_{md} = 1.5\ \mathrm{m\ s^{-1}} \quad \text{and} \quad v_{mp} = 1.33\ \mathrm{m\ s^{-1}}$$

5.7 THERMAL SOARING

The conditions under which thermals form most readily are mainly to be found in hot climates, but even in temperate zones thermals can occur wherever the ground is covered by regions of disparate vegetation. The substantial vertical components of velocity in a thermal are ideal for maintaining aloft birds, particularly raptors such as vultures, eagles, and buzzards, which have developed methods of exploiting these conditions. In order to stay within the field of influence of a given thermal a bird must modify its normal gliding flight, in a straight line, to a constant circling motion, which is the essence of thermal soaring. For a bird to maintain height, the updraught velocity of the thermal must just balance the sinking speed of the bird relative to the air. If the updraught velocity exceeds the sinking speed, then of course the bird is able to climb to still higher altitudes.

When a bird circles at constant altitude it must bank or tilt so that a component of the aerodynamic lift is used to balance the centrifugal force generated by the rotational motion.

The relationships between the updraught velocity, v, the velocity of the bird relative to the air, V_a, and the velocity of the bird relative to the ground, V_g, required to maintain the soaring motion in a horizontal plane, are illustrated in Fig. 5.11.

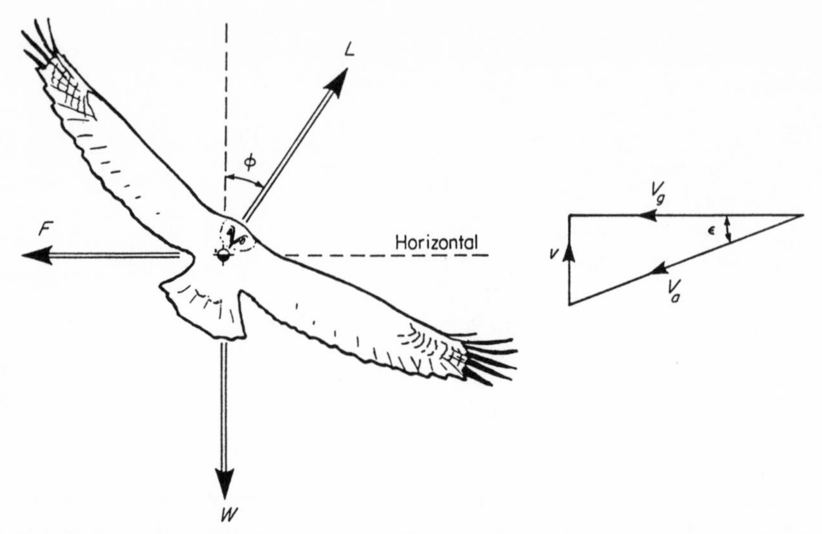

Figure 5.11. Bird in horizontal turning flight. The angle of bank and the glide angle are denoted by ϕ and ε respectively.

If the bird is moving in a circle of radius, r, then the centrifugal force, F, is given by

$$F = \frac{mV_g^2}{r}$$

where m is the mass of the bird.

For equilibrium in the horizontal direction

$$\frac{mV_g^2}{r} = L \sin \phi \tag{5.18}$$

where ϕ denotes the angle of bank.

For equilibrium in the vertical direction, the vertical component of the lift, $L \cos \phi$, must just balance the weight, mg. Thus:

$$W = mg = L \cos \phi \tag{5.19}$$

From equations (5.18) and (5.19) the angle of bank, ϕ, is given by:

$$\tan \phi = \frac{V_g^2}{rg} \tag{5.20}$$

and the lift, L, is:

$$L = mg\left(1 + \left(\frac{V_g^2}{rg}\right)^2\right)^{1/2} \tag{5.21}$$

Equation (5.21) demonstrates that, when moving in a circular path, a bird must generate aerodynamic lift in excess of that required under conditions of motion in a straight line. The required lift force increases as the radius describing the circular motion decreases, or as the groundspeed V_g increases.

The aerodynamic lift and drag are, as for normal gliding flight, determined by the velocity of the bird relative to the air, V_a. Hence, the corresponding aerodynamic equations developed for normal gliding flight can be carried over and applied to thermal soaring.

Careful observation in regions where a variety of different birds soar on the thermals, such as the Indian and central African plains, shows that they take to the wing one after the other in a systematic hierarchy. The smallest birds, with the lowest wing loadings, are first, and the last on the wing are the largest birds, with the highest wing loadings. These observations (Hankin, 1913) can be explained as follows.

To be able to exploit a thermal to gain height, a bird must have a sinking speed which is less than the updraught velocity. In the early morning updraught velocities are low but increase progressively during the course of the day as the ground is heated by the sun. It is to be expected from our earlier analysis that the birds which will have the lowest sinking speeds will be those with the lowest minimum drag speeds, that is birds with low wing loadings. Birds with higher wing loadings will have to wait until the vertical convection currents have become sufficiently vigorous to sustain their more exacting demands upon the thermals.

Figure 5.12. Long distances across country can be covered by the combination of alternate periods of thermal soaring and gliding.

By alternating between periods of thermal soaring and gliding, various species of birds have discovered a most efficient method of making long-distance flights across country (Pennycuick, 1972). A bird gains height in a thermal and, having reached a sufficient altitude, leaves it to embark on a period of gliding flight towards another thermal. During this phase the bird loses height but, once having reached a new thermal, is then in a position to gradually regain altitude once more. Vultures in central Africa are known to use this technique and, when rearing young, will traverse distances well over 100 km (62 miles) from their nests in search of carrion. This flight strategy is also employed during migration by white storks (*Ciconia ciconia*) and certain migratory eagles.

The wings of birds specializing in thermal or slope soaring, such as the vultures (*Gyps*), buzzards (*Buteo*), eagles (*Aquila*), or the marabou stork (*Leptoptilos crumeniferus*), are typically rectangular in shape and of aspect ratio between 6 and 8. A special feature of these rectangular wings is the disposition of the primary feathers which, when the bird is soaring, are capable of being separated and splayed out along the wing tip. Clearly the ability of the bird to use its feathers in this manner confers upon it certain aerodynamic advantages, but the way in which this advantage is derived is still incompletely understood. It appears that one of the benefits is in the form of a lower vortex drag, and this explanation is supported by the fact that a test aircraft fitted with wing tip sails, similar in conception to the emarginated feathers of vultures, has achieved useful reductions in this component of drag. At the tip of a normal aircraft or bird wing there is a substantial upwash velocity component associated with the trailing vortex system, and this velocity component combines with the freestream velocity to increase the effective angle of incidence of the airstream in this region. The wing tip sails on the test aircraft were carefully designed (Spillman, 1978) to exploit this local effect, which tilts the lift generated by the sails forward, effectively providing a small component of thrust. It needs no great stretch of the imagination to deduce that the aerodynamic forces on the splayed primaries of soaring birds probably operate in a similar manner, the net effect being an overall reduction of the vortex drag of the wing. Another possible advantage conferred by the separated primaries is that, when the bird is flying below the minimum drag speed, the necessary control of the flight speed can be effectively achieved by the adjustment of groups of feathers rather than of the wing as a whole.

5.8 DYNAMIC SOARING

The albatrosses (Diamedeidae) have evolved a soaring technique, known as dynamic soaring, which is entirely their own, involving flight in the layers of air immediately above the surface of the sea. Dynamic soaring rests not on the presence of vertical components of velocity such as are found in thermals, but on an entirely different principle dependent upon the appreciable gradients of horizontal wind speed to be found at these low levels. In effect, the wind behaves like a boundary layer. Close to the surface of the sea the windspeed is low and increases rapidly with height above the surface. Typically albatrosses operate in the layers of wind up to about 12 m (36 ft) above the sea.

In discussing the flight behaviour of an albatross it is important to distinguish between the velocity of the bird relative to the air, V_a, and its velocity relative to a

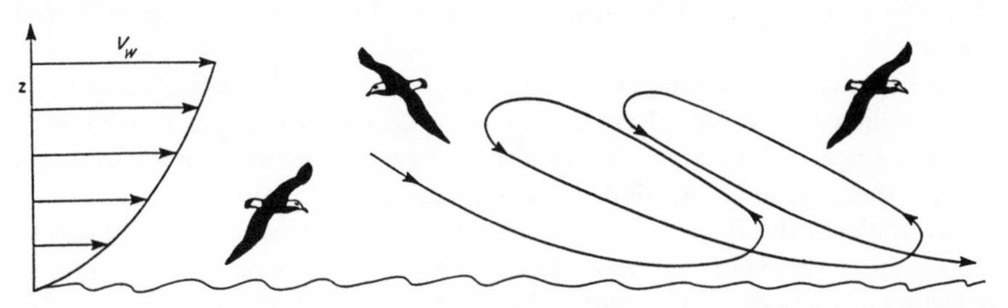

Figure 5.13. Dynamic soaring of the albatross. The wind profile is indicated on the left, and the flight path of the albatross, as seen by an observer at rest, is shown schematically on the right.

stationary frame of reference, V_g. The former we shall call the airspeed, whilst the latter will be referred to as groundspeed. The windspeed is denoted by V_w.

There are two separate phases to the flight pattern of the albatross (Fig. 5.13). First of all, the bird glides directly into the wind gaining height but losing groundspeed as a result of the combined effects of drag and gravity. A stage comes when the bird reaches the peak of its climb and it then turns through 180° and faces downwind to start the second phase. In this phase the albatross loses height and rapidly gains groundspeed. When the bird approaches close to the surface of the sea it again turns and faces upwind to repeat the first phase of the flight pattern. This process of rising and falling is repeated in endless sequence, the albatross being swept along in the direction of the prevailing wind. It hardly ever needs to flap its wings and so it manages to remain airborne with a minimum expenditure of energy.

In order to sustain itself in the air the albatross requires open sea and the presence of prevailing winds which do not sweep the bird on to land. Such ideal conditions exist deep in the southern hemisphere, where the albatross is able to circulate around the South Pole, maintaining itself within a defined latitude band, which ensures that it will not encounter land. The geography of the northern hemisphere, with its large land masses, renders it unsuitable for the albatrosses, and explains their absence—except for rare visitors which have lost their bearings—in this part of the world.

During the first phase of its flight pattern the albatross loses groundspeed, and hence kinetic energy, whilst at the same time it gains height, and hence potential energy. When the wind shear becomes insufficient for its purpose the albatross turns downwind. Relative to the ground, it accelerates rapidly, the loss in potential energy providing the source of the increase in kinetic energy. However, as it rises and falls, the albatross is all the time affected by aerodynamic drag which is constantly acting to deplete its energy level. In the absence of any other factor this tendency, over a number of cycles of the flight pattern, would result in a reduction in the kinetic energy level. Eventually the airspeed would drop below the stalling speed and the bird would no longer be able to sustain itself in the air. That this does not happen is explained by the fact that the albatross is able to extract energy from the wind sufficient to counteract the energy being dissipated due to drag.

The explanation is as follows. For convenience, a moving frame of reference is selected such that the local velocity of the wind is zero. Thus, relative to a stationary frame of reference, our chosen reference frame constantly changes speed in the horizontal direction as the albatross rises and falls.

The motion of the albatross is three-dimensional in character for, in addition to gaining and losing height, and flying up and down wind, on occasions as it loops to change directions it flies across the wind. Hence, in our chosen frame of reference, the total velocity of the albatross is the airspeed, V_a, which has components u_a, v_a and w_a in the horizontal, vertical, and transverse directions respectively. We take u_a and v_a to be positive when the bird moves in the same direction as the wind and when it climbs, respectively. In our moving frame of reference, at any instant the kinetic energy of the albatross is $\frac{1}{2} mV_a^2$ and its potential energy is mgz, where m is its mass, g is the acceleration due to gravity, and z is the height above the sea. Due to the air resistance the albatross is expending energy at a rate given by the product of drag and airspeed. We have noted already that, relative to a stationary frame, our chosen frame of reference is constantly changing speed in the horizontal direction; in other words it has a horizontal component of acceleration. The magnitude of this acceleration is deduced in the following way. In a small interval of time, δt, the albatross climbs through a height, δz, given by the product $v_a \times \delta t$. In this interval of time the horizontal velocity of the frame of reference must change by an amount $(\mathrm{d}V_w/\mathrm{d}z) \times \delta z$. The horizontal acceleration, a, is equal to the rate of change of horizontal velocity. Thus we have

$$a = \frac{\delta z}{\delta t} \times \frac{\mathrm{d}V_w}{\mathrm{d}z} = v_a \frac{\mathrm{d}V_w}{\mathrm{d}z}$$

and this term is positive when the bird climbs. The quantity $\mathrm{d}V_w/\mathrm{d}z$ is, of course, the wind gradient: the rate of change of windspeed with height. Because of the acceleration of the chosen frame of reference, there is a related inertial force, the effect of which must be taken into account in the energy balance. The inertial force acts in the opposite direction to the acceleration vector and is given by the product $-ma$. The rate of working of this inertial force is given by the product $(-ma) \times u_a$; that is:

$$-mu_a v_a \frac{\mathrm{d}V_w}{\mathrm{d}z}$$

At any instant in time the energy equation for the albatross expresses the fact that the rate of change of kinetic energy plus potential energy is equal to the rate of working of the inertial force, less the rate of energy depletion due to aerodynamic drag.

The quantities m and $\mathrm{d}V_w/\mathrm{d}z$ are always positive. During the climb into the wind u_a is negative and v_a is positive, whereas during the descent u_a is positive and v_a is negative. Hence, throughout most of the flight pattern the product $u_a v_a$ is negative. Whilst this condition is satisfied, the rate of working of the inertial force is positive. In other words the albatross is extracting energy from the sheared wind profile. Only as it turns to change direction does the albatross go through short phases where the inertial force acts to further deplete the energy level. Inevitable variations exist between the trajectory and flight mechanics adopted by the albatross during one cycle and the next. However, taking the average over several cycles of the flight pattern, provided the mean rate of working of the inertial force is sufficient to at least balance the loss of energy due to aerodynamic drag, the sum of kinetic energy and potential energy suffers no overall depletion.

The albatross has an aspect ratio of about 18 compared with an aspect ratio of about 6 or 7 for the vulture, a typical bird using the thermal soaring strategy. In contrast, the wing loading of the albatross is approximately double that of the vulture: 140 N m^{-2} (2.8 lbf/ft^2) compared to 70 N m^{-2} (1.4 lbf/ft^2). As equation (5.9) demonstrates, this ensures that the albatross has a reasonably high gliding speed. This is an important requirement because when the albatross has completed its climbing phase and first turns downwind its airspeed is at a minimum. A high gliding speed ensures that stall does not occur at this critical stage in the flight pattern. Here, whilst the wing is operating at near its maximum lift coefficient, a further benefit of the wing geometry is manifested: the very high aspect ratio ensures that the lift-dependent drag is not excessive.

Although it was convenient to discuss the mathematical analysis as though the surface of the sea was level, in practice the shearing action of the winds causes an appreciable swell. Under these conditions the albatross uses slope soaring techniques close to the waves to augment its more general use of dynamic soaring.

Chapter 6

Bird Flight: II—Flapping Flight

In the design of conventional aircraft the fixed wing serves the purpose of weight support whilst thrust generation is provided by a jet engine, or by a propellor plus power plant. Thus, the two roles of weight support and thrust generation are independent of each other.

With the flight styles of birds so far considered, flight speed was maintained either by exploiting gravity or by extracting energy from the wind. There was no need for thrust generation, and aerodynamic lift could be developed with the wings held in a fixed configuration.

The flight styles which we are about to discuss—fast forward flight, slow forward flight, and hovering—all involve flapping of the wings and are, from an aerodynamic viewpoint, extremely complex. In fast forward flight the flapping motion of the wings is used principally for thrust generation, whereas in hovering the beating of the wings is entirely devoted to weight support. Slow forward flight represents a mode of flight intermediate between these two extremes, and the flapping contributes both to weight support and thrust generation.

6.1 THE AERODYNAMICS OF THE WING IN FAST FORWARD FLIGHT

At this stage it is appropriate to focus attention on the wing motion required to promote forward flight at normal cruising speeds. Under these conditions, as a good approximation, the processes of lift and thrust generation can be separated (Lighthill, 1974), as shown schematically in Fig. 6.1, where the motion of a wing during a complete cycle is shown. For the generation of a constant lift force, a constant angle of incidence is required as the wing is flapped.

For the generation of thrust the wing must be twisted, being pitched down during the downstroke and in the opposite direction during the upstroke. The combined motion is shown on the right of Fig. 6.1, where relative to the bird, the

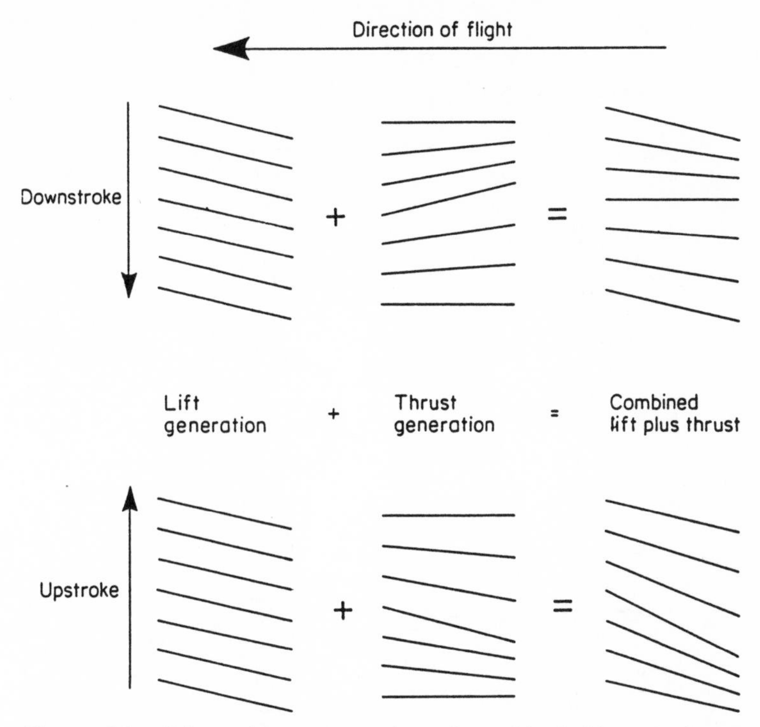

Figure 6.1. Schematic representation of combined lift and thrust generation of a flapping wing in normal horizontal flight. Each line represents the chord of the wing section at some arbitrary position across the span of the wing.

wing moves essentially in a single vertical plane. During much of the downstroke the wing is approximately parallel to the direction of flight, but during the upstroke the wing assumes a strongly pitched-up orientation. This pattern of wing movement is observed in the flight of gulls and other birds.

The air forces producing thrust oppose the wing movement during both the downstroke and the upstroke, whereas in generating lift the air forces oppose the wing in the downstroke but assist the movement during the upstroke. As a consequence, the wing is very heavily loaded during the downstroke, which explains the large size of the pectoralis-major muscles responsible for the wing's movement during that phase of the flapping cycle and the much smaller size of the supracoracoideus muscle which acts during the upstroke. Greenewalt (1975) has analysed the musculature of many birds, whose body mass ranges from 5 g to 10 kg. Throughout this entire range the mass of the pectoralis-major muscle was typically 15 per cent of the body mass; the mass of the minor pectoral muscle was typically 1.5 per cent of the body mass.

An examination (Lighthill, 1977) of the forces acting on a flapping wing shows that in addition to the zero-lift drag and lift-dependent drag terms identical to

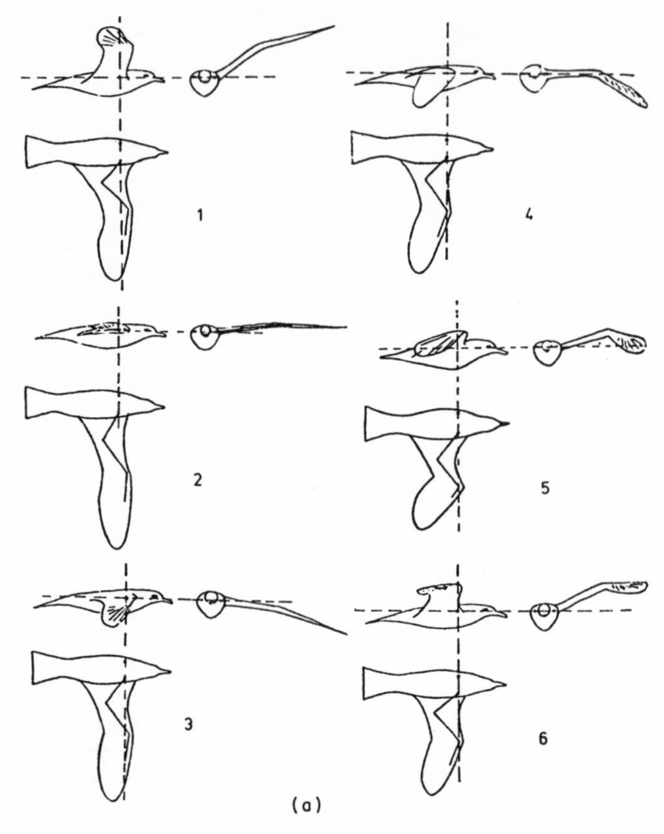

Figure 6.2(a). Wing movements of a gull in normal horizontal flight. (1) Beginning of downstroke; (2) middle of downstroke; (3) approaching end of downstroke; (4) beginning of upstroke; (5) middle of upstroke; (6) approaching end of upstroke. (Reproduced, by permission of the Company of Biologists Ltd, from Brown, (1953), *J. Exp. Biol.*, **30**, 90–103.)

those for a fixed wing, there is an additional vortex drag term resulting from the unsteady flapping motion.

The above simple explanation of the aerodynamics of a flapping wing is appropriate to fast forward flight. The lift is generated in essentially the same manner as in flight styles where the wings are held fixed; the flapping movement of the wing is almost entirely devoted to the generation of thrust.

The foregoing description is, of course, somewhat idealized. Due to anatomical factors there is much greater scope for rotational and translational components of wing movement towards the wing tips, in contrast to the restricted motion available over the inboard section of the wing. As a consequence, the region of wing occupied by the primary feathers is particularly involved in thrust production, whereas that part of the wing to which the secondary feathers are attached makes the main contribution to the generation of aerodynamic lift.

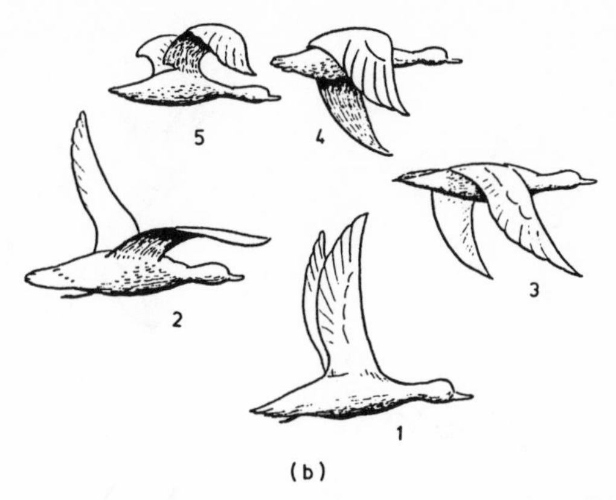

Figure 6.2(b). Wing movements of a duck in normal horizontal flight. (1) Beginning of the downstroke; (2) part-way through the downstroke; (3) end of the downstroke; (4) and (5) part-way through the upstroke. (Reproduced, by permission of the Macmillan Publishing Company, from Hylander, C. J. (1959), *Feathers and Flight*. Copyright © 1959 by Macmillan Publishing Company.)

6.2 THE AERODYNAMICS OF THE WING IN SLOW FORWARD FLIGHT

On a number of occasions reference has been made to the maximum lift coefficient that can be generated by a fixed wing. This determines, for the fixed wing, the minimum speed, the stalling speed, V_s, below which insufficient aerodynamic lift can be generated to sustain flight. Because a bird is free to move its wing in both horizontal and vertical planes, we must modify our ideas very considerably to understand the aerodynamics of a flapped wing in slow forward flight, at the reduced speeds adopted just prior to landing, for example. In fact many birds are capable of remaining airborne at forward speeds substantially below the speed at which they would stall if they maintained a fixed-wing configuration. We can illustrate this point in the following way.

Consider first of all a bird flying forward at a speed, V, just above its stalling speed in a fixed-wing configuration. Now, in flapping its wings the bird superposes, relative to its centre of mass, a forward and backward movement to augment the upward and downward movement of the wings. Relative to the bird, denote the forward velocity of the wing by u and the downward velocity by w.

The magnitude and direction of u and w vary with spanwise position along the wing and throughout the flapping cycle. At any instant the effective angle of incidence at any point on the wing, that is the angle between the chord line and the incident airstream, is equal to the geometric incidence plus an additional term $w/(V + u)$ due to the motion of the wing. Thus, the wing is tending to operate at higher incidences and hence at higher values of C_L, providing the stalled condition is avoided, as it moves forward and downward. The lift

Figure 6.3. Wing movements of a pigeon in slow forward flight. This flight style is adopted just after take-off or prior to landing. (Reproduced, by permission of the Company of Biologists, from Brown, (1948), *J. Exp. Biol.*, **25**, 322–333.)

produced is proportional to the square of the relative velocity, that is $(V + u)^2 + w^2$. Averaged over the whole of the flapping cycle, during which u and w take on both positive and negative values, the quantity $\overline{(V + u)^2 + w^2}$ exceeds V^2, so that the lift is enhanced by the increased airspeed resulting from the movement of the wing. Detailed mathematical analysis of the flapping wing (Lighthill, 1977), which is too complicated to be reproduced here, indicates that there is a further factor augmenting the lift at slow forward speeds, when the velocities u and w become comparable in magnitude to V. The direction of the thrust vector, perpendicular to the stroke plane of the wing, in addition to providing a component to overcome drag also makes a significant contribution to weight support which more than offsets a slight reduction in the contribution from the lift vector which itself is rotated out of the vertical plane. The enhancement of the

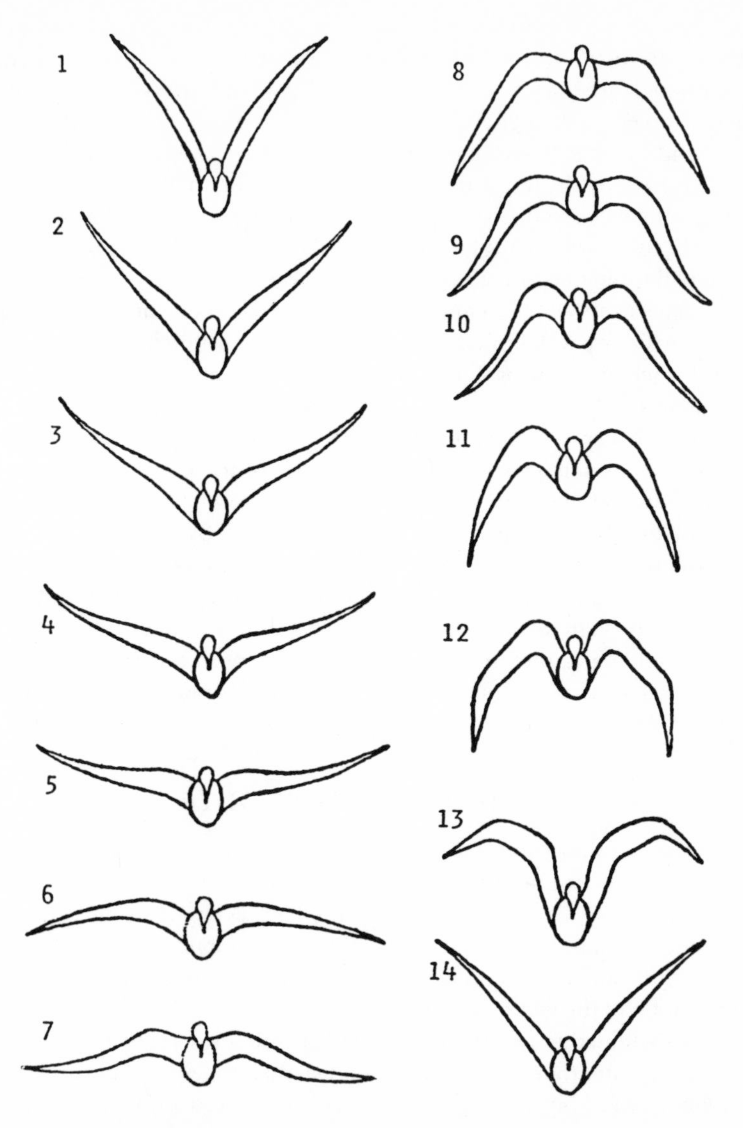

Figure 6.4. Wing movements of a gull in slow flapping flight, viewed from the front. The substantial bending of the elbow and wrist joints during the up-stroke is evident. (From Horton-Smith, (1938) *The Flight of Birds*, H. F. and G. Witherby, London.)

relative airspeed and the canting of the thrust vector together explain why birds are able to fly forward at speeds well below their basic fixed-wing stalling speed. This mode of flight is adopted by many small birds just prior to landing.

In straight level flight at fast forward speeds the motion of the wing is, relative to the bird, essentially in a vertical plane. As the flight speed is progressively reduced, so that the stalling speed of the fixed wing is eventually approached, the bird must introduce forward and backward components to the wing's motion, and increasingly the wing is bent at the elbow and wrist joints during the upstroke. At very low flight speeds the wing flapping becomes increasingly vigorous and, as the hovering condition is approached, weight support is provided entirely by the large-amplitude flapping motion of the wing.

Most small garden and woodland birds having wings of modest aspect ratios can cover the full flight regime from fast forward flight to hovering by utilizing the appropriate flapping action of the wing. However, many larger birds, and birds with high aspect ratio wings, cannot flap their wings sufficiently vigorously, due to anatomical or strength limitations, and so cannot sustain flight at low forward speeds. For such birds the problems presented at take-off and landing are severe, as we shall presently see.

At low forward speeds the alula or bastard wing plays an important role. It works in much the same way as a leading edge slot used on the wing of certain modern aircraft, helping to preserve attached flow over the main wing to angles of incidence higher than would be possible in its absence. By delaying separation in this way the wing can operate effectively at higher angles of incidence. Higher values of maximum lift coefficient are thereby attained and so the bird is better able to control its flight at these low forward speeds. Tests have revealed that the action of the alula is automatic; it is drawn out into position by the large suction pressures created in its vicinity when the wing is operating at a high angle of incidence.

6.3 HORIZONTAL FLIGHT

In order to sustain horizontal forward flight in still air a bird must generate a component of thrust sufficient to overcome the aerodynamic drag to which it is subject. To achieve this it is necessary for the bird to flap its wings. At once the aerodynamics of flight become considerably more complicated, for in birds both weight support and thrust generation are simultaneously provided by the complex motion of the wing.

Consider sustained horizontal flight through still air. For the flight trajectory to remain horizontal the aerodynamic lift must at all times be equal and opposite to the weight of the bird. For a steady horizontal speed to be maintained, the thrust, generated by the wing alone, must just balance the drag, part of which is due to the wing, the remainder being contributed by the bird's body. Since a bird has to flap its wings up and down to produce the thrust, the actual thrust generated varies appreciably during the flapping cycle. As a result, a bird cannot hold a precisely constant forward speed in flapping flight. It accelerates forward slightly during those parts of the wingbeat cycle when thrust is produced, and decelerates during the remainder of the cycle. Hence, with each cycle of the wing, the instantaneous flight speed oscillates slightly about the average flight speed, and as a consequence the lift and drag also oscillate about mean values. However, these variations are normally very small and only when we discuss the detailed aerodynamics of the flapping wing need we be concerned with the essentially unsteady nature of the problem; in considering the general aerodynamic performance of birds in sustained horizontal flight we shall assume a constant average flight speed as the basis of our discussion, with no significant loss in accuracy.

6.4 CHARACTERISTIC FLIGHT SPEED

For a bird in steady horizontal forward flight at a constant speed, V, through still air, the aerodynamic lift, L, and weight, W, are in balance so that

$$W = mg = \frac{1}{2} \rho V^2 S C_L = L \tag{6.1}$$

where ρ is the air density, S is the bird's wing area and C_L is the mean lift coefficient. This simple equation contains a number of important results. Firstly, a bird can to some extent control its flight speed by controlling the mean lift coefficient at which it operates; the higher the lift coefficient, the lower the flight speed, and vice-versa. Secondly, a reduction in air density, such as occurs with increase in altitude, requires an increase in flight speed for sustained flight. Finally, birds with low wing loading, W/S, are able to maintain height at low flight speeds and vice-versa.

In fast horizontal flight the aerodynamic drag can again be expressed by equation (5.4) provided we ignore a small additional component of vortex drag due to the flapping motion of the wing. The minimum drag speed, V_{md}, defined in equation (5.5) is the speed at which the bird can fly horizontally a given distance with lowest energy expenditure. The minimum power speed, V_{mp} ($= 0.76\ V_{md}$) represents the optimum flight speed for a bird wishing to remain airborne at constant altitude for a given period of time with a minimum of energy expenditure. These conditions will be considered further when migration is discussed.

As the wing is responsible for the entire thrust generated in flight, yet equally contributes substantially to the aerodynamic drag, we are faced with a dilemma: there is no precise means of determining the magnitudes of the thrust and overall drag of a bird in fast flapping flight although approximate estimates can be made from a knowledge of the gliding performance. Consequently, when observations are made of the typical flight speeds of birds, it is not a simple matter to interpret the observed cruising speed, V_c, in terms of V_{md} or V_{mp}. Mathematical analysis shows that, in normal flight over short distances, a bird would derive aerodynamic advantage by selecting a cruising speed close to V_{mp}, whereas during migratory flight a higher cruising speed nearer V_{md} is appropriate. These considerations are certainly borne out by the observation that birds generally migrate at speeds above those they employ under general flight conditions (Meinertzhagen, 1955).

Since the characteristic flight speeds, V_c, V_{md}, and V_{mp}, are all proportional to the square root of wing loading, birds with low wing loadings fly slower than those which are more highly loaded. This explains the contrast between, for example, the slow, casual relaxed flight style adopted by the crow (*Corvus*) and the much more urgent approach typical of ducks (*Anas*) and pigeons (*Columba*).

As has already been demonstrated in the context of non-flapping flight, scaling laws provide useful insights into the flight characteristics of birds. Assuming geometrical scaling laws can be applied, then the mass, m, of a bird is proportional to the cube of a characteristic length, l. This length scale could be the wing span or the distance from beak to tail; it is not a matter of importance since, with the assumption of geometrical similarity, all the dimensions are in fixed proportions. Similarly, the wing area, S, is proportional to l^2. Thus, according to these scaling laws, the wing loading, mg/S, is directly proportional to l or to $m^{1/3}$. Greenewalt (1962, 1975), has checked the dimensions of a large number of different birds, ranging in size from the smallest hummingbirds to the largest birds capable of taking to the air. Geometrical similarity is most evident between birds within groups having similar life styles, such as the duck family, or raptors, or woodpeckers, etc. Within each family the scaling laws are closely representative of the bird's characteristics. Furthermore, taking all the birds together, the scaling law provides a good indication of the general trend of increased wing loading with mass, with a mean relationship being given by

$$\frac{W}{S} m^{-1/3} = 90 \text{ N m}^{-2} \text{ kg}^{-1/3}$$

There are, of course, variations about the mean correlation, reflecting the different flight styles of different types of bird. For example, typical values (in $\text{N m}^{-2} \text{ kg}^{-1/3}$) for garden birds are between 50 and 100, whereas raptors have a lower range of values, from about 30 to 60 reflecting their need to carry prey, whilst many aquatic birds, such as ducks, have values in excess of 100.

Based on steady-state aerodynamics, the maximum lift coefficient that a bird can normally achieve is between 1.5 and 2. For any bird, the minimum flight speed it can sustain in horizontal flight, that is its stalling speed, V_s, can then be estimated from a knowledge of its wing loading using equation (5.8). From our earlier discussion of scaling laws, it follows that V_s is proportional to $l^{1/2}$, or to $m^{1/6}$. The velocities V_{md} and V_{mp}, as well as V_s, all depend upon $(W/S)^{1/2}$ and hence, according to the scaling laws, vary as $l^{1/2}$ or $m^{1/6}$. Greenewalt has investigated the law between flight speed V and m for various groups of birds including falcons (Falconiformes), herons (Ardeidae), owls (Strigiformes), woodpeckers (Picidae), wrens (Troglodytidae), ducks (Anatinae), coots (Fulicinae), divers (Gaviidae), grebes (Podicipediformes), pigeons (Columbidae), swans (Cygninae), etc. and has shown that measurements in the field are in close conformity with the predictions based on geometrical scaling laws. There is, therefore, very considerable evidence that the characteristic flight speeds of birds, as exemplified by V_s, V_{md}, and V_{mp} tend to increase with the mass, m, and the simple geometrical scaling law $V \propto m^{1/6}$ provides a good approximation of the general relationship.

6.5 FLIGHT CAPABILITY AND THE INFLUENCE OF WEIGHT

The minimum power required to sustain horizontal flight is given by the product of drag and velocity, evaluated at the minimum power speed. At this condition it can be shown that the lift-dependent drag is three times as great as the zero-lift drag. Thus, the minimum power, P_{min}, required to overcome air resistance can be expressed in the form

$$P_{min} = 2\rho V_{mp}^3 S C_{D0}$$

Using the scaling laws, $V_{mp} \propto 1^{1/2}$, $S \propto 1^2$ we deduce

$$P_{min} \propto l^{7/2} \quad \text{or} \quad P_{min} \propto m^{7/6} \tag{6.2}$$

These scaling laws are equally valid at the minimum drag speed, V_{md}. Since the power required to fly at the minimum drag speed is $P = \rho V_{md} S C_{D0}$, the relations expressed by (6.2) follow as before.

The important result inherent in (6.2) is the demonstration that the power requirements for sustained flight increase more rapidly than does the mass of the bird. The arguments regarding the availability of power from the bird's muscles are as follows. There seems to be an upper limit to the sustained rate of working of muscles, and this upper limit is about 220 Watts for each kg of muscle. The proportions of a bird's weight taken up by muscles is, with the exception of hummingbirds (Trochilidae), typically one part in five, from which it is deduced that the maximum power that could be sustained in horizontal flapping flight is about 44 m Watts, where m is the bird's mass.

However the power output of 220 Watts per kg of muscle can only be delivered if the muscles are contracted rapidly and repeatedly at frequencies of above about 10 cycles per second. For most birds the wingbeat frequency, which tends to diminish with increase in body mass, is less than this figure, and the work that can be done in a single contraction becomes the effective limiting factor on available power. The power output is therefore even lower than would be possible at higher frequencies. The inevitable consequence of the conflicting trends of required and available power is that there is a maximum weight for birds above which sustained flight is not possible.

It is important to appreciate that the factors we have been discussing provide a useful qualitative insight into overall trends, but they cannot be pressed to yield specific quantitative data. The reason is as follows. The principle of geometrical similarity has been invoked. But, as we all know, birds come in a great variety of shapes. It is vital for every species of animal, including birds, to find its own ecological niche in order to survive, and so evolutionary pressures encourage diversity, which is the direct antithesis of similarity. Indeed, although it is not profitable for us to investigate such matters in detail, the scaling laws we have been considering pinpoint where evolutionary changes could appear in order to confer specific aerodynamic advantages. The survivors amongst the bird kingdom are survivors precisely because where they have attained special skills or advantages of any sort, these have not compromised the many other factors—including the ability to avoid predators, the ability to raise offspring in sufficient numbers, etc.—which ultimately ensure that one generation is succeeded by another.

With all the host of complex factors influencing evolutionary development, it is interesting to note that the mass of the heaviest flying birds is approximately 12 kg and this limit is approached in four different orders of birds, all with quite distinct life styles (Pennycuick, 1972). We list the heaviest member of each order: the California condor (*Gymnogyps californianus*), the Kori bustard (*Ardeotis kori*), the white pelican (*Pelecanus onocrotalus*), and the mute swan (*Cygnus olor*). By way of contrast we note that specimens of the heaviest flightless bird, the ostrich (*Struthio camelus camelus*), have been recorded at 156 kg (345 lb).

6.6 HIGH-SPEED FLIGHT

Of considerable interest is the question of the fastest bird on the wing in horizontal flight. By common consent the swifts (Apodidae) hold this record. But in fact an accurate answer to this question is a matter of some difficulty. Whereas there is no problem in deriving accurate statistics on the groundspeeds of different birds, it is a very different matter to assemble data on airspeeds. The problem arises because, to obtain the airspeed from a knowledge of the groundspeed, the wind speed and its direction must be known. Since wind speed varies with altitude and is subject to unsteadiness due to gusts, meteorological data on mean wind speeds are insufficiently accurate to make the required adjustments. From the available evidence (Meinertzhagen, 1955), it appears that a number of birds, including hummingbirds (Trochilidae), mallards (*Anas platyrhyncas*), and racing pigeons (*Columba palumbus*), in addition to the swifts, are capable of high airspeeds—in excess of 25 m s^{-1} (56 mph).

In Russia, during 1942, a spine-tailed swift (*Chaetura caudacuta*) was reliably measured to have a groundspeed of 47.5 m s^{-1} (106 mph), the highest value recorded. Whilst many naturalists have celebrated the virtuoso performances of the swift on the wing, however, it is still not possible at present to give an accurate figure of the maximum airspeed it can attain in horizontal flight.

The main factor in the attainment of high flight speed is the capacity to develop sufficient thrust to overcome the high aerodynamic drag which is an inevitable feature of such flight conditions. It seems that the swifts are particularly well adapted to the generation of thrust in flight.

In swifts the mass of the flight muscles as a proportion of body mass is significantly higher than the average figure for birds as a whole. Not only is this true of the main flight muscle, the pectoralis-major, which powers the downstroke, but it applies even more forcibly to the smaller muscle powering the upstroke. Typically amongst birds the masses of these two muscles are in the ratio of ten to one; in the swift the increased mass of the muscle powering the upstroke reduces this ratio to five to one. Thus the swift is capable of substantial thrust generation in both the down and upstrokes. This musculature is used in conjunction with the lunate or crescent-shaped planform of the wing to produce thrust by essentially rigid-body movement of the wing, since bending of the wrist joint is eliminated.

The swift is distinguished also by its exceptional wingbeat frequency. Data on wingbeat frequency have been assembled for many birds, based on direct measurements in the field. However, such is the extreme rapidity of the wing movement of the swift that estimates based on direct visual observations are hopelessly imprecise. Accurate figures, it would seem, await the analysis of evidence collected by the cine or television camera.

The wing movements of birds may be described in terms of a frequency parameter $\omega c/V$, where V is the flight speed, c is the wing mean chord, and ω is the wingbeat frequency,* in radians per second. Values of between 0 and 0.5 are representative of most birds, but the swift has a significantly higher value, of about 1. Lighthill (1974) has suggested that it is for values of the frequency parameter of this order that the lunate shape becomes particularly efficient as a thrust-generating surface.

In summary, high-speed flight in the swift can be attributed to four factors—musculature, the lunate wing shape, rigid-body movement of the wing, and the exceptional wingbeat frequency—all of which contribute to efficient thrust generation.

Whilst the morphology of the swift is primarily determined by considerations of thrust production, the other side of the coin—drag reduction—is not entirely neglected. Swifts have forked tails, and whereas the tail is widely spread to provide lateral control at low speeds, it is closed in fast flight, so reducing the surface area and with it skin-friction drag, generally giving the body a streamlined shape ideally suited to high-speed flight.

Assuming that the geometrical scaling laws are applicable, further deductions can be made concerning flight speed. For geometrically similar birds $V \propto l^{1/2}$, and so it follows that the largest birds are the fastest. Amongst the swifts the longest wing span is shared by the alpine swift (*Apus melba*) and the larger black swifts (*Cypseloides*), and so it is probable that the fastest of all birds is to be found from amongst their number.

6.7 HOVERING FLIGHT

The hummingbirds (Trochilidae) are renowned for their ability to hover, whilst carefully removing the nectar from flowers. Hovering, that is maintaining a zero groundspeed in still air, is a flight style common to insects, but in the bird world as a sustained form of flight it is unique to hummingbirds, although many small birds are capable of hovering for short periods.

During hovering the hummingbird's wing tips describe a figure of eight, and substantial aerodynamic forces are generated during both the down and upstrokes, which are essentially symmetrical. This type of wing motion is referred to as normal hovering. It is quite different from the wing motion employed by other hovering birds. In contrast to the hummingbird, those other birds which briefly engage in hovering use the wing motions of slow forward flight (section 6.2), so that almost all the useful work is done during the downstroke; the upstroke is very lightly loaded and makes little contribution to weight support. This type of wing motion is called avian hovering.

To hover, a hummingbird does not have to overcome the aerodynamic drag

* The wingbeat frequency, ω, in radians per second, is equal to $2\pi\omega_c$, where ω_c is the frequency in cycles per second.

associated with forward flight but must concentrate all its efforts in sustaining its own weight in the air. For this purpose the stroke plane of the wing must be approximately horizontal, whilst the body is held nearly vertical. The movements of the wing are similar to those of the moth in hovering flight, which is illustrated in the next chapter.

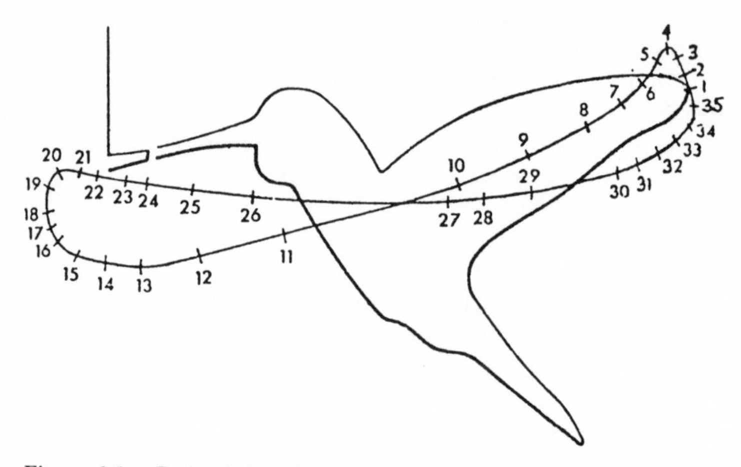

Figure 6.5. Path of the wing tip of a hummingbird during hovering flight. Successive positions are shown at equal increments of time. (Reproduced, by permission of Verlag für Wissenschaft und Forschung, from Stolpe and Zimmer, (1939), *Der Vogelflug*.)

The flow of air through the stroke plane in hovering flight is due solely to the induced downwash velocity, whereas for a bird in forward flight the bird's own airspeed contributes to the flow through the inclined stroke plane.

Some simple sums indicate the magnitudes involved. The flapping wings may be considered to operate over a circular disc within the stroke plane with a diameter equal to the wingspan of the bird. The air influenced by the wing may be viewed as a sort of jet which is accelerated from rest, reaching half its final velocity at the stroke plane, the process of acceleration to the downwash velocity, u, assumed constant across the jet, only being completed some distance below the stroke plane. Defining the area of the disc swept out by the flapping wings by $A_D(=\pi b^2/4)$, the area of the jet when the uniform downwash velocity, u, is achieved is, from the continuity principle $A_D/2$, and the rate of mass flow through the jet is $\rho A_D u/2$.

Considering the forces acting in the vertical direction, the weight of the hovering bird must be just balanced by a lift force given by the rate of change of momentum of the air in the vertical direction. Thus

$$mg = \rho\frac{A_D u^2}{2} \tag{6.3}$$

The induced power, P, required to generate the kinetic energy in the downwash is

$$P = \frac{1}{2}\rho\frac{A_D u^3}{2} \tag{6.4}$$

Elimination of u between equations (6.3) and (6.4) yields

$$P = \frac{1}{2}mg\left(\frac{2mg}{\rho A_D}\right)^{1/2} \tag{6.5}$$

This simple analysis, which is in fact a further example of the application of actuator disc theory, includes a number of assumptions all of which lead to the underestimate of the required power. Thus, the effective area over which the wings actually operate is reduced by the presence of the body. Furthermore, the assumed constant distribution of downwash velocity is the most favourable that could be achieved; in practice the velocity distribution is inevitably non-uniform. Also this simple theory assumes that the downwash is continuously produced by the wing, whereas in reality the flow generated by the beating wing is unsteady.

Despite these detailed reservations about the precision of the actuator disc theory, equation (6.5) can be expected to give a reasonable estimate of the power required for hovering. As an example, we shall consider the performance of the large hummingbird *Patagona gigas*, which has a mass of about 20 g and a span b = 30 cm. Taking $\rho = 1.21$ kg m^{-3} and $g = 9.81$ m s^{-2} we deduce from equation (6.5) that to hover the hummingbird must expend energy at a rate in excess of 0.23 Watts. An approximate check on this figure can be obtained by estimating the available power. Hummingbirds have a ratio of flight muscle mass to body mass of about 30 per cent and, having wingbeat frequencies typically between 15 and 50 cycles per second (well in excess of 10 cycles per second), are able to generate a continuous power output of about 220 Watts per kilogram of muscle. Thus the available power for *Patagona gigas* is estimated at about 1.32 Watts, comfortably exceeding the required power.

The pied flycatcher (*Ficedula hypoleuca*) is an example of a bird which utilizes brief periods of hovering flight when it is on the wing. When its hovering performance is interpreted (Norberg, 1974) in terms of steady-state aerodynamics, calculations show that the flycatcher generates a mean lift coefficient of 5.3, a figure far in excess of values typical of the aerodynamics of fixed-wing aircraft. Such calculations indicate that the interpretation of hovering flight in terms of steady-state aerodynamics is unrealistic, and explanations which take account of the inherently unsteady nature of the motion must be sought (see section 6.13).

The hovering flight style of the hummingbird is quite distinct from the characteristic hovering technique adopted by certain birds of prey. The kestrel (*Falco tinnunculus*), or wind hover as it is sometimes called, and the buzzard (*Buteo buteo*) are capable of maintaining a zero groundspeed, but cannot do this in still air. The kestrel and buzzard operate in a wind of velocity V_w; they achieve a zero groundspeed by facing directly into the wind and flying into it with an airspeed V_w. This style of flight is perhaps most appropriately described as wind-assisted hovering.

During wind-assisted hovering the bird employs the wing motions of slow forward flight, as previously discussed in section 6.2.

6.8 TAKE-OFF AND LANDING

For many birds the transition from resting on the ground or a branch of a tree to sustained flight at speed is a smooth and seemingly uneventful process. After a firm push-off with the legs and a few beats of the wing, the result is a rapid increase in the flying speed, and the bird is on its way. Similarly, when such birds come into land, they lose flying speed, perhaps augmenting the control exercised by the wing by spreading their tail and extending their feet to increase drag, pass through a phase of slow forward flight and come safely to rest. But for some birds the take-off and landing phases of flying are agonizing and embarrassing. Paradoxically, this predicament is most extreme for some of the supreme flyers, birds which once on the wing are majestic masters of the air. Examples in this category are the swifts (Apodidae) and albatrosses (Diomedeidae).

Flight at low forward speeds is aerodynamically inefficient and highly costly from the point of view of energy expenditure, so that in the take-off process a bird is aiming to move from rest to a flight speed close to the minimum power speed, V_{mp}, with the minimum of effort. The main problems of take-off and landing occur amongst those birds which cannot utilize the lift generation techniques of vigorous flapping at slow forward flight speeds discussed earlier, either because they are unable to generate sufficient power or because their wings are not suited to this mode of flight. With large, heavy birds, the power requirement is limiting. It will be evident also that low aspect ratio wings are more suited to these flight regimes, as they present fewer anatomical problems, being better able to withstand the appreciable forces and bending moments developed, and from an aerodynamic viewpoint they are also more efficient. Thus the birds which experience difficulty at take-off and landing include amongst their numbers the heaviest flying birds, birds with high wing loadings and birds with high aspect ratio wings.

Such birds depend primarily upon conventional, fixed-wing aerodynamics for lift, and flapping motion is used largely for the generation of thrust. In order to sustain flight the bird must develop an airspeed equal to or in excess of the stalling speed, V_s, corresponding to the maximum lift coefficient, $C_{L\max}$, in order to generate a lift, L, equal in magnitude to its weight, W.
We have

$$W = L = \frac{1}{2} \rho V_s^2 S C_{L\max}$$

so that to remain airborne the flight speed, V, must exceed V_s. Thus,

$$V > \left(\frac{2W}{\rho S C_{L\max}}\right)^{1/2}$$

a relation which clearly shows how the minimum flight speed increases with wing loading.

The need to build a nest and raise a family is one of the main reasons all birds have to come to terms with the necessity to land. Many birds which find take-off and landing difficult choose to nest at considerable heights above the ground. Sea birds, for example, use cliff faces, swifts nest in church towers or other high buildings, and some birds build exposed nests in the tops of trees. The reason for this is as follows. A particularly convenient means of building up airspeed is to trade off potential energy in order to increase kinetic energy. Hence, in leaving their nests, these birds follow a descending path and thereby quickly attain an airspeed sufficient to generate aerodynamic forces with which they assume mastery of the air. When coming in to land, by approaching from below, these birds can reach their nests at low speeds, exchanging their kinetic energy for potential energy.

For those birds which spend some of their life on the ground, a long take-off run is required and so a flat surface, free of restrictive vegetation is required. The Kori bustard (*Ardeotis kori*) runs along the ground on long legs, flapping its wings, building up speed before it is able to take to the air. Amongst aquatic birds, the mute swan (*Cygnus olor*), in taking off from the water, flaps its wings vigorously and use its webbed feet to assist itself forward. A considerable expanse of water is required before the bird can obtain sufficient forward speed with which to generate the necessary aerodynamic lift which allows it to leave the water, from which it climbs slowly away. In such large birds the power which the flight muscles can provide is only marginally in excess of the minimum required for powered flight, and so sustained flight is only possible over a small range of speeds, around the minimum power speed.

6.9 INTERMITTENT FLIGHT STYLES

A mechanism which some birds employ is to alternate two flight styles, the objective being an overall reduction in fuel consumption or some related benefit (Rayner, 1977; Ward-Smith, 1984).

In adopting a bounding flight style a bird flaps its wings in the powered phase, expending rather more power than is required solely for fast level flight and this excess is used to gain a small amount of height. The bird then folds its wings, thereby reducing air resistance, moves through the air following a ballistic trajectory like a projectile, and in the absence of significant lift, starts to lose height. After a short period it flaps its wings again, regains height, and repeats the flight pattern, so that overall its flight path oscillates about a mean horizontal trajectory (Fig. 6.6). The increase in energy expenditure in the flapping phase is offset against the saving achieved in the second phase. Examples of birds using this bounding flight style are the finches (Fringillidae) and woodpeckers (Picidae), of which the latter are representative of the largest birds using it.

A simple mathematical analysis of the bounding flight style may be made in the following way. The powered flight phase is represented by motion at a constant velocity V along a circular arc of radius R, and the ballistic trajectory is assumed to follow a parabolic path. It is assumed that the bird changes from flapping flight to unpowered flight at the point where the actual flight path crosses the horizontal mean flight path. The angle between the flight path and the horizontal at this point is denoted by α.

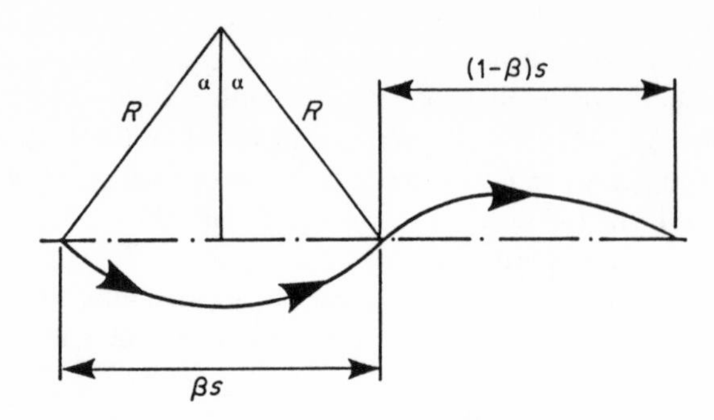

Figure 6.6. Schematic representation of the flight path during bounding flight.

The horizontal distances traversed in the powered and unpowered phases are denoted by βs and $(1 - \beta)s$ respectively.

By considering the geometry of the powered phase the relation

$$\sin \alpha = \frac{\beta s}{2R} \tag{6.6}$$

is obtained.

During the powered phase, due to the motion along a circular arc, the bird experiences a centrifugal force of magnitude mV^2/R. This force acts normal to the flight path but, to a first approximation, it can be taken to be vertical. A force balance then shows that for equilibrium the aerodynamic lift generated by the bird must equal the sum of the weight plus centrifugal force. Thus

$$L = mg\left(1 + \frac{V^2}{Rg}\right) \tag{6.7}$$

From equation (5.5), the drag D is given by

$$D = \frac{1}{2}\rho V^2 S C_{D0} + \frac{KL^2}{(\pi/2)\rho V^2 SA} \tag{6.8}$$

Writing

$$K_1 = \frac{1}{2}\rho S C_{D0} \tag{6.9}$$

and

$$K_2 = \frac{K(mg)^2}{(\pi/2)\rho SA} \tag{6.10}$$

and combining equations (6.8), (6.9), and (6.10) we obtain

$$D = K_1 V^2 + \frac{K_2}{V^2}\left(1 + \frac{V^2}{gR}\right)^2 \tag{6.11}$$

Thus the lift-dependent drag is augmented as a result of the motion along the circular flight path.

We turn now to the ballistic phase. If the lift and drag are negligible the flight path is parabolic, and the distance traversed in this phase is then given by the exact relation

$$(1 - \beta)s = \frac{2V^2 \sin\alpha \cos\alpha}{g} \tag{6.12}$$

Significant simplifications can be made to the preceding equations by assuming α to be small, so that $\sin\alpha \approx \alpha$ and $\cos\alpha \approx 1$. Introducing these approximations, equations (6.6) and (6.12) may be substituted in equation (6.11) to yield

$$D = K_1 V^2 + \frac{K_2}{\beta^2 V^2} \tag{6.13}$$

In bounding flight, the total work done in travelling a distance s is equal to the work done during powered flight over the distance βs, since it is assumed that no external work is done during the ballistic phase.

Thus the work done W_D is travelling a distance s is

$$W_D = \left(K_1 V^2 + \frac{K_2}{\beta^2 V^2}\right)\beta s \tag{6.14}$$

which is a minimum when β and V are related by the expression

$$\beta = \frac{1}{V^2}\left(\frac{K_2}{K_1}\right)^{1/2} \tag{6.15}$$

Writing V_{md} to represent the minimum drag speed for sustained horizontal flight, equation (6.15) can be rewritten as

$$\beta = \frac{V_{md}^2}{V^2} \tag{6.16}$$

For speeds V below the minimum drag speed, β is greater than 1, and it follows that bounding flight does not lead to energy savings at these low speeds. As a strategy for reducing fuel consumption, bounding flight is seen to be restricted to speeds in excess of V_{md}.

Substitution of equation (6.15) in equation (6.14), shows that, for all values of $V > V_{md}$, the work done W_D in travelling a distance s in bounding flight is identical to the work done during sustained horizontal flight at the minimum drag speed. Basically, bounding flight is a strategy for increasing flight speed without incurring the normal penalty of increased energy expenditure, associated with sustained horizontal flight.

However this benefit can, in practice, only be won over a restricted range of values of β (or V).

The power required for flight at velocity V is given by

$$P = DV = K_1 V^3 + \frac{K_2}{\beta^2 V} \tag{6.17}$$

Substitution of equations (6.15) and (6.16) in (6.17) yields

$$P = \left(\frac{V}{V_{md}}\right)^3 [2K_1^{1/4} K_2^{3/4}] \tag{6.18}$$

Thus, using P_l to denote the continuous power output required to sustain steady horizontal flight at the minimum drag speed and P_{bf} to denote the power required in the flapping phase of bounding flight we have, from equation (6.18), the relation

$$P_{bf} = \left(\frac{V}{V_{md}}\right)^3 P_l = \beta^{-3/2} P_l$$

P_{bf} increases rapidly with decrease in β below 1, and it is this consideration which places the main limitation on the extent to which the bounding flight style can be exploited.

Observations of the siskin (*Carduelis spinus*) have shown that it alternates periods of flapping flight which last for about 0.25 s with periods of unpowered flight of about 0.35 s duration. Taking $\beta = (0.25/0.6) = 0.42$, we calculate $V = 1.55\ V_{md}$ and $P_{bf} = 3.72\ P_l$ for flight under these conditions.

A distinct flight pattern, known as undulating flight, is adopted by heavier birds, such as gulls (Laridae) and crows (Corvidae), in which a period of conventional gliding flight is followed by a few powered strokes of the wing during which height is regained (Fig. 6.7).

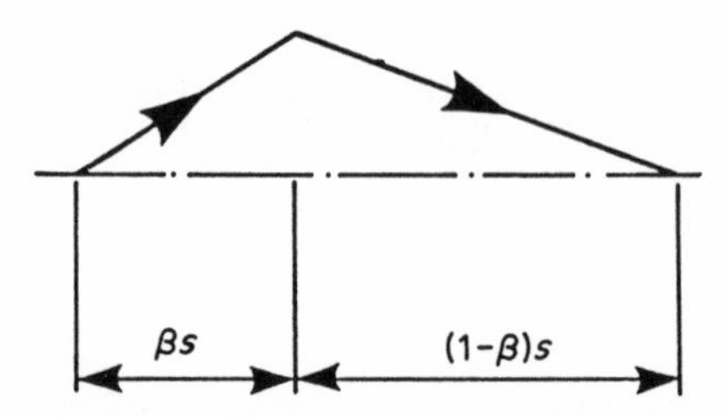

Figure 6.7. Schematic representation of the flight path during undulating flight.

To fly a distance s in straight and level flight with minimum energy expenditure a bird will fly at the minimum drag speed (equation (5.5) with the weight mg substituted for L).

Now consider the alternative strategy. Suppose that the bird glides over a horizontal distance $(1 - \beta)s$, and climbs under powered flight over a horizontal distance βs. The angle of the glide and climb relative to the horizontal are ε_1 and ε_2, respectively. The geometry of the flight path is such that

$$(1 - \beta)s \tan \varepsilon_1 = \beta s \tan \varepsilon_2$$

During the glide and climb the lift L is given by mg cos ε_1 and mg cos ε_2, respectively.

The speed which gives the minimum glide angle is the minimum drag speed, which is related to the glide angle by tan $\varepsilon_1 = D/L$.

During the climb the bird must do work to overcome drag (which is again minimized by flight at the minimum drag speed) and to regain the potential energy lost during the glide.

In the glide and climb the lift L, and consequently the lift-dependent drag, are less than that in straight and level flight. Analysis based on equation (5.4) shows that, if D^l_{md} is the magnitude of the drag at the minimum drag speed in level flight, then the minimum drag in the glide and climb are, respectively, $D^l_{md} \cos \varepsilon_1$ and $D^l_{md} \cos \varepsilon_2$.

The work done in the climb is given by

$$D^l_{md} \cos \varepsilon_2 \times \frac{\beta s}{\cos \varepsilon_2} + mg\,\beta s \tan \varepsilon_2$$

where the first term is the work done against drag and the second is the increase in potential energy. Now

$$\tan \varepsilon_2 = \frac{(1-\beta)}{\beta} \tan \varepsilon_1 = \frac{(1-\beta)}{\beta}\left(\frac{D}{L}\right)_{\text{glide}} = \frac{(1-\beta)}{\beta}\,\frac{D^l_{md}}{mg}$$

so that we deduce that the work done during the climb is given by $D^l_{md}s$, which is exactly the same as the energy expended in horizontal flight over the distance s.

The selection of this flight style is therefore not a matter of reducing the external work done, and so its explanation must be sought elsewhere.

Using the symbol P_{uf} to denote the power required in the flapping phase of undulating flight and P_l to denote the continuous power output for steady level flight at the minimum drag speed, we have

$$P_{uf} = \frac{P_l}{\beta}$$

so that the principal feature of undulating flight is neither a change of flight speed nor of external work done but rather a change to the maximum rate at which energy is expended during the course of the flight.

The pectoral muscles of birds are primarily composed of fast-contracting oxidative, glycolytic fibres. Laboratory tests show that there is an optimum rate of contraction at which these muscles produce maximum power, and that there is a sharp decline in power production, and system efficiency, at other contraction rates. Consequently, it is disadvantageous for a bird to regulate its flight performance by varying wingbeat frequency, and, for a given species of bird, within a defined tolerance band the flapping frequency of the wings bears a fixed relation to the optimum contraction rate of the pectoral muscles. Normally, therefore, flight performance is regulated by varying the force generated during a stroke of the wings according to the number of individual muscle fibres that are recruited, or by varying the amplitude of the stroke, or by varying wing geometry.

The flight styles of undulating and bounding flight can be seen to provide an alternative strategy. The musculature of the small and medium-sized birds using these flight styles is determined primarily by the stringent conditions at take-off, and for these birds the continuous power output, P_l, required to sustain steady level flight at the minimum drag speed is lower, typically by a factor of two or more, than the maximum available power, P_a. By compressing the power requirements into a proportion β of the distance travelled, a bird is able to fly at an intermittent power loading P_l/β^n, which allows the musculature to operate at or near peak efficiency. For undulating flight $n = 1$ and, because the external work done is independent of the choice of β, the value of β can be selected solely on the basis of optimizing the efficiency of performance of the musculature. For bounding flight $n = 3/2$ and both the consideration of flight speed and the efficiency of the musculature influence the optimum choice of β.

6.10 MIGRATORY FLIGHT

Amongst the most prodigious flying achievements of the birds are the feats of endurance and distance traversed during migration.

The arctic tern (*Sterna paradisaea*), in particular, covers extraordinary distances during its migratory flights. This species nests along the northern coasts of the American, Asian, and European continents, and then flies from these arctic regions far south across the equator. From the evidence of ringed birds, it has been established that distances of over 12,000 km (8000 miles) are not uncommon. The greatest distance that a bird has been recorded away from its nest site is some 19,000 km (12,000 miles); this distance was attained by an arctic tern ringed in July 1955 at the Kandalaksha sanctuary on the White Sea, USSR, and subsequently caught alive in May 1956 near Fremantle, Western Australia. This particular ringing gives no indication of the speed at which this migratory journey was made. Here the evidence of other bandings is relevant. An arctic tern ringed as a nestling in July 1928 in Labrador, Canada, was recovered dead 116 days later in Natal, South Africa, a destination some 13,000 km (8500 miles) from the point of departure. Making due allowance for the time which elapsed before the bird left its nesting area, an average speed of migration of approximately 150 km per day (100 miles per day), sustained over 85 days, is indicated.

Consistent with deductions we made earlier, it has been observed that birds have a general tendency to migrate at flight speeds above their normal cruising speeds. Over land their journey is interspersed with periods during which they alight for food and rest. An average speed of migration of 150 km per day (100 miles per day) is not untypical of numerous species of migrating birds. Whilst they benefit from a tail wind, birds are not averse to migrating into a head wind, and results (Eastwood, 1967) using radar tracking techniques reveal that they are capable of maintaining a constant heading during a period when the direction of the wind changes.

To sustain itself in the air during migration a bird must constantly do work against the resistance of the air. Power must be supplied continuously to the flight muscles, and the source of this power is the energy provided by the burning of fuel, that is the oxidization of the body fats derived from the food it has consumed.

In preparation for its long flight a bird must take a considerable amount of fuel on board, by eating food in excess of its normal consumption; in flight the body weight of the bird continuously diminishes as this fuel is burnt up.

We can analyse the situation in the following way. Let the energy content of the body fat per unit mass be denoted by e and the efficiency of energy conversion by η. Then the energy required to overcome drag in a small time interval δt must be supplied by the oxidation of a mass of fat δm. Thus

$$\eta e \delta m = P\delta t = DV\delta t \tag{6.19}$$

and the burning of this fat leads to a reduction in the weight of the bird, W, given by

$$\delta W = -g\delta m \tag{6.20}$$

whilst the bird travels a horizontal distance, δR, given by

$$\delta R = V\delta t \tag{6.21}$$

Allowing the time interval δt to become vanishingly small, equations (6.19) to (6.21) can be combined to give the differential equation

$$\mathrm{d}R = -\frac{e\eta}{gD}\mathrm{d}W \tag{6.22}$$

In level flight $L = W$, so equation (6.22) can be rewritten as

$$\mathrm{d}R = -\frac{e\eta}{g}\frac{L}{D}\frac{\mathrm{d}W}{W} \tag{6.23}$$

which, for constant L/D, integrates to

$$R = \frac{e\eta}{g}\frac{L}{D}\log_e\frac{W_1}{W_2} \tag{6.24}$$

where R is the range covered (i.e. the horizontal distance flown), and W_1 and W_2 are respectively the weights of the bird at the start and finish of the flight.

A representative figure for the amount of work which can be obtained from unit mass of fat ($e\eta$) is 8×10^6 Joules per kilogram. Taking $g = 9.81\ \mathrm{m\ s^{-2}}$, equation (6.24) can be evaluated to determine R for given values of L/D and W_1/W_2.

For a given quantity of body fat consumed in flight (i.e. constant W_1/W_2), it follows from equation (6.24) that the maximum range is achieved for flight at the maximum value of L/D. Our earlier analysis has shown that this condition is attained at the minimum drag speed, V_{md}, for which

$$\left(\frac{L}{D}\right)_{\max} = \frac{1}{2}\left(\frac{\pi A}{KC_{D0}}\right)^{1/2} \tag{6.25}$$

An alternative interpretation of equation (6.24) shows that, for a given range, the minimum consumption of body fat occurs for flight at V_{md}.

For L/D = constant, C_L is also constant, and the general equation for lift, evaluated at $V = V_{md}$,

$$L = W = \frac{1}{2} \rho V_{md}^2 S C_L \tag{6.26}$$

shows that with W decreasing throughout the flight, the product ρV_{md}^2 must decrease accordingly. Since

$$V_{md} = \left(\frac{K}{\pi A C_{D0}}\right)^{1/4} \left(\frac{2W}{\rho S}\right)^{1/2} \tag{6.27}$$

a bird can adopt one of two flight strategies, as inspection of equations (6.26) and (6.27) shows.

At constant altitude the density ρ is constant, and the first strategy is to fly at constant altitude, allowing the speed V_{md} to diminish as $W^{1/2}$. An alternative strategy is to maintain a constant value of V_{md}, but to increase altitude in such a way that W is directly proportional to ρ. Whichever strategy is adopted the overall consumption of fuel is the same. However there is one advantage of flight at increased altitude: the higher value of V_{md} means that the duration of the flight is reduced. This is an important factor motivating birds to fly at increased altitude during migration.

The existence of strong winds can have a significant effect on migratory strategy and Pennycuick (1972) has calculated some of the effects. For small birds, with low wing loadings and low minimum drag speeds, there may be advantages in exploiting the earth's boundary layer to minimize the effects of the winds, and sightings of migratory flights close to the surface of the sea are probably explained by the existence of unfavourable winds at altitude.

Calculations based on equation (6.24) show that if a bird, with a typical lift/drag ratio of 6, were to fly non-stop for about 3400 km (2200 miles) its body weight at the end of the flight would be only half that at the start. With the same lift/drag ratio, a range of 2000 km (1300 miles) would result in a final weight two-thirds of the initial weight. These simple estimates show that the range depends upon the proportion of body weight required as fuel, but is independent of the actual weight of the bird. Indirectly, they also serve to emphasize the importance which environmentalists place upon conserving the 'staging posts' at which birds stop for refuelling on the migratory routes.

The majority of migratory flights are made at modest altitudes (Eastwood, 1967), typically below about 1200 m (4000 ft) with a preferred altitude band from 600 m to 900 m (2000 ft to 3000 ft). However, there are many records of flights at much higher altitudes and, over sea, at very low levels. Using radar techniques, migratory flights of flocks of birds, thought to be shore birds, have been recorded in the USA at altitudes as high as 6000 m (20,000 ft), and recordings at 5000 m (16,000 ft) have been obtained over England. In mountainous regions of the globe the lie of the land forces migrating birds to ascend to high altitudes. Some of the migratory routes through the Himalayan Mountains, for example, are along passes at altitudes as high as 5500 m (18,000 ft).

On occasions, sightings from aircraft have allowed the bird species in migratory flight to be positively identified (Tyne and Berger, 1959). Rooks (*Corvus frugilegus*) have been recorded at 3300 m (11,000 ft), geese at 2700 m (9000 ft), and lapwing (*Vanellus vanellus*) at 2600 m (8500 ft). At the opposite extreme, some species of bird have been seen skimming within 9 m (30 ft) of the sea during migratory flight, examples being turtle doves (*Streptopelia turtur*) and ortolan buntings (*Emberiza hortulana*).

6.11 BIRD FLIGHT AT ALTITUDE

In discussing migratory flight, some examples have already been given of the altitudes attained by birds. In mountainous regions many birds live permanently at high altitude. For example, Tibet is on average some 5000 m (16,400 ft) above sea level. Nevertheless, over 500 species have been recorded within its borders, and of these about 400 probably breed there (Vaurie, 1972).

At high altitudes birds encounter two difficulties affecting flight not met at lower levels. These are a depletion in oxygen, essential to the burning of its fuel, and a reduction in air density, which diminishes the aerodynamic lift force the bird can sustain at a given flight speed. For example, at 5000 m (16,400 ft) the atmospheric density is only 63 per cent of the value at sea level and so, to generate the same amount of lift in horizontal flight, a bird needs to fly 26 per cent faster and the power consumed in doing this increases by the same 26 per cent. These penalties in horizontal flight are severe, and to remain airborne for any length of time at very high altitude it would seem essential for a bird to exploit the upcurrents on the sides of the mountains.

Many birds which are skilled soarers have been recorded at high altitude, but the fact remains that numerous other species which do not resort to this mode of flight have been observed as well. Some of the most valuable observations made at high altitude are those of A. F. R. Wollaston and R. W. G. Hingston, members of the British expeditions on Mount Everest in 1922 and 1924, and knowledgeable ornithologists. The following birds were recorded by them (Hingston, 1927) at altitudes above 6100 m (20,000 ft) and probably constitute records for the species:

Richard's pipit (*Anthus novaeseelandiae*)	6100 m (20,000 ft)
Black redstart (*Phoenicurus ochruros*)	6100 m (20,000 ft)
Blue hill pigeon (*Columba rupestris*)	6100 m (20,000 ft)
Jungle crow (*Corvus macrorhynchos*)	6400 m (21,000 ft)
Raven (*Corvus corax*)	6400 m (21,000 ft)
Hoopoe (*Upupa epops*)	6400 m (21,000 ft)
Alpine accentor (*Prunella collaris*)	6400 m (21,000 ft)
Bearded vulture or Lammergeyer (*Gypaetus barbatus*)	7300 m (24,000 ft)
Alpine chough (*Pyrrhocorax graculus*)	8200 m (27,000 ft)

The altitude attained by the Alpine chough, several of which attached themselves to the expedition party in 1924, is generally recognized as the highest recorded for bird flight.

6.12 FORMATION FLYING

In discussing the aerodynamics of a single bird in flight we have recognized that the passage of the bird through the air cannot be effected without the air itself being caused to move. In particular, the generation of lift in fast forward flight is associated with the appearance behind the wing of a pair of trailing vortices. Within each vortex the flow is rotating in such a sense that, for a single bird, between the wing tips a downwash velocity occurs, whereas outside the wing tips an upwash velocity component results. In steady level flight the air suffers no overall change of momentum in the horizontal direction as the thrust and drag forces are in equilibrium, whereas the rate of change of momentum of the air in the vertical direction must just balance the weight of the bird.

Birds often take to the wing in groups (Hummel, 1983). For example, some species of birds come together in large flocks before roosting or during long migratory flights. In such large flocks the birds are so numerous that, to the untrained eye, the position of one bird relative to another within the flock looks disorganized and random. Some birds adopt a much more systematic formation when flying as a group. The supreme example is the precise V-formation, assumed by skeins of geese and on occasion by other birds, such as ducks, pelicans, and cranes. A variation on this same theme is the echelon formation taken up by the same group of birds, as well as by gulls and perhaps others besides.

The question arises: can birds adopt particular flight formations which confer certain aerodynamic advantages? As we shall see, the answer is: yes.

First of all it is instructive to note that there is one formation that is not preferred. Birds do not fly one behind the other, line astern. It is worth pausing to consider why this should be so. After all, racing cyclists (and motorcyclists and car drivers) have all discovered that there is a positive advantage to be gained by tucking in behind someone else. This technique is often referred to as slipstreaming. The advantage of this formation arises because the air resistance of the leading rider, his profile drag, is manifested in the form of a wake which is pulled along behind him. In this wake, the air is moving forward in the same direction as the rider but at a slower speed. Anyone riding within the influence of this wake has a lower effective airspeed and consequently is subject to a reduced aerodynamic drag. Hence, in a team event, all of the riders ride close together line astern behind the leader and, with the exception of the leader, have the advantage of a lower air resistance than if they were riding separately. As the leading rider is subject to the greatest resistance he has to work the hardest and so the riders in the formation continually change positions, taking it in turn to be the leader.

It might be thought that these arguments are equally applicable to birds in

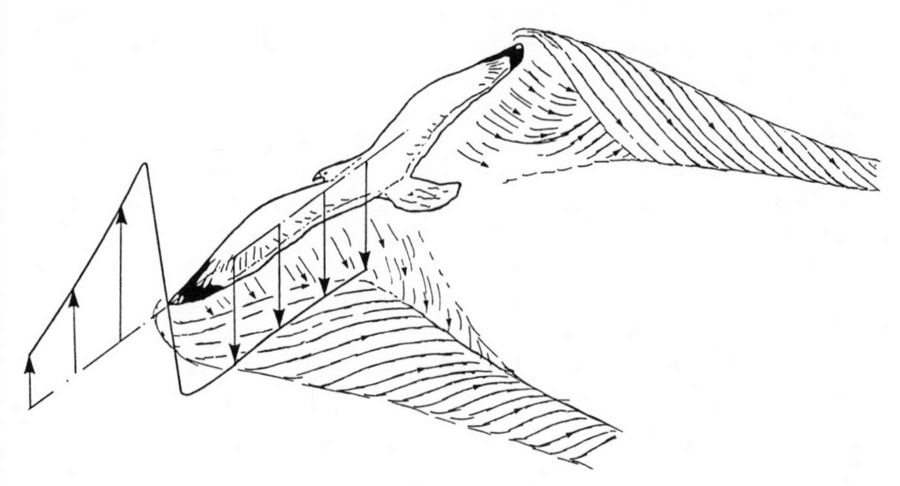

Figure 6.8. Schematic representation of the flow field associated with the trailing vortex system of a bird. On the left wing the downwash and upwash velocity components are illustrated for a typical plane through the wing. The process of rolling-up of the downwash field to form a pair of trailing vortices is shown on both wings.

flight. They also are subject to profile drag, so any reduction in this would imply that the wing would need to generate a smaller thrust component. But there is a vital difference between a bird in flight and a rider moving over the ground. In generating thrust and drag simultaneously by aerodynamic means a bird causes no overall change to the momentum of the air in the horizontal direction, so that the large wake exploited by the slipstreaming cycle rider is not available for birds to use. There is another vital difference between the bird and rider.

Behind a bird is a downwash field, which is not present behind the rider. The lift and drag forces acting upon a second bird flying in the downwash field are inclined from the vertical and horizontal directions respectively. In consequence, the second bird must generate increased lift and thrust to compensate for these particular effects. We must conclude that certain species of birds have confirmed for themselves the existence of these extra demands placed upon them by the downwash field of other birds; it is equally evident that these same birds have discovered the aerodynamic advantages to be gained by exploiting the upwash velocity components beyond the span of the wing.

A useful starting point from which to elucidate the aerodynamic advantages of particular flight formations is to consider a group of birds flying in the same horizontal plane, in a wing tip to wing tip formation. For simplicity we assume that we can base our discussion on fixed-wing aerodynamics; we ignore the component of vortex drag due to the flapping motion of the birds' wings. Each of the birds in the formation derives the advantage of flying in the upwash field of its neighbours and as a result at a given flight speed, V, the vortex drag of each bird and the vortex drag of the formation as a whole is reduced. The profile drag of the formation remains unchanged. To obtain the full benefit of the reduction in the vortex drag it is advantageous for the formation to fly at an airspeed substantially below the optimum for a single bird, at which the profile drag and vortex drag terms are equal.

It has been calculated (Lissaman and Shollenberger, 1970) for 25 birds flying in this horizontal line-abreast formation, that the optimum cruising speed for the formation is 24 per cent below that for a single bird and, at this speed, a range 71 per cent greater than that for a lone bird can be achieved for the same overall energy consumption.

There is a very useful theorem of classical aerofoil theory, known as Munk's stagger theorem, which allows the results for wing tip to wing tip formation to be applied to other formations. This theorem states that any member of an assembly of lifting surfaces may be translated to a new position by moving it in the direction of its flight path without altering the total vortex drag of the system.

The really important result arising from Munk's theorem, so far as the overall performance of the formation is concerned, is that the fore-and-aft positions of the individual birds are of secondary importance; the vital factor is their positions when all the birds are projected on to a single plane normal to the direction of flight. Any formation which yields a wing tip to wing tip line-abreast configuration is optimal. Any other projection is inferior. A line-abreast projection with an occasional gap, whilst not as efficient as it could be, is at least aerodynamically advantageous, but line-abreast projections with overlapping birds, or formations involving the flight of birds at different heights, could mean that the birds are interacting unfavourably from an aerodynamic viewpoint.

In the line-abreast formation the birds in the centre of the formation derive a greater aerodynamic advantage than those near the outer fringes. By continuously changing their positions in the fore-and-aft direction the birds are able to redistribute the benefits conferred on each individual by the overall formation. Also by interchanging positions throughout a long flight all the birds can make an equal contribution to the overall work load required of the formation. This is largely what happens when birds fly in echelon formation. But a more sophisticated means of achieving the same result is to use a formation pattern which automatically equalizes the aerodynamic advantages amongst the birds whilst all of them maintain a constant position within the formation. Mathematical analysis shows that this is closely approximated by a V-shaped formation. This formation has a further decisive advantage; it provides all birds with direct visual contact with their neighbours, so allowing the overall pattern of the formation to be readily established and maintained.

Birds flying in V-formation usually number less than one hundred at a time. The number of birds in a large flock can run into several thousands (May, 1979), and in these circumstances the birds are clearly prepared to trade off some of the aerodynamic advantages of flight formation for other reasons; for example, proximity to the centre of gravity of the flock. For small birds this is a useful survival strategy.

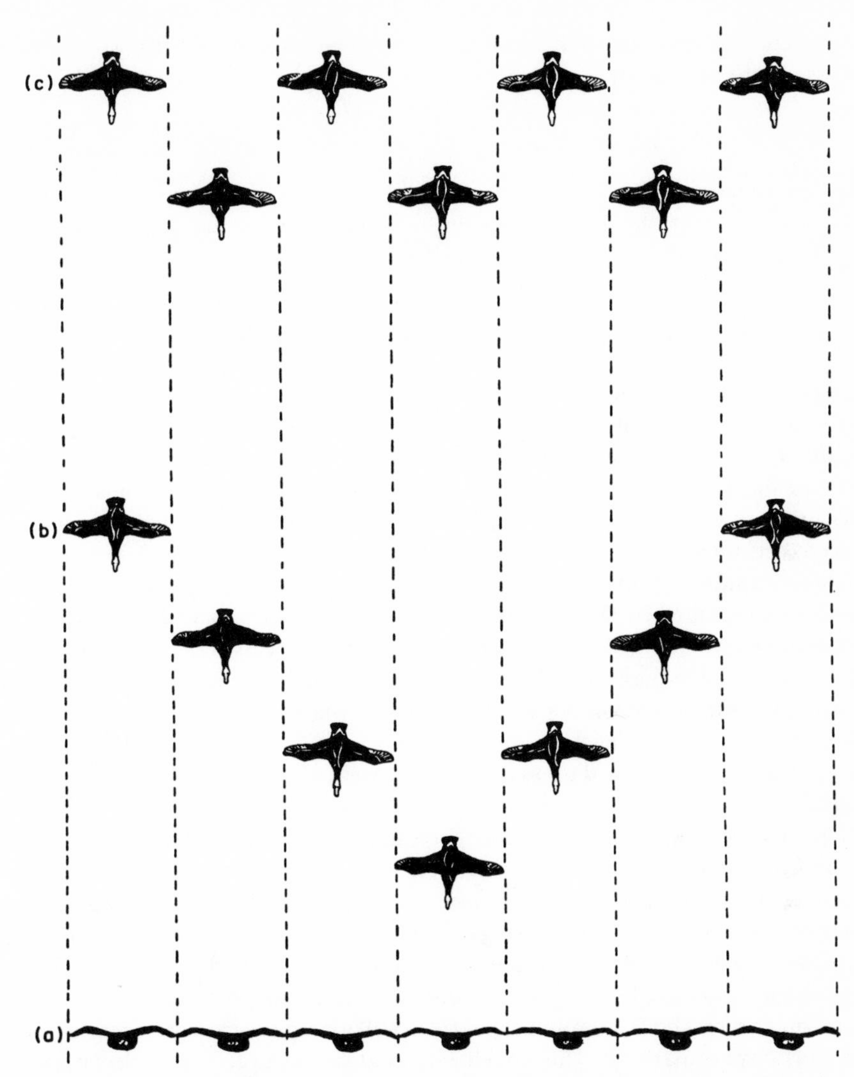

Figure 6.9. Flight formations in which the overall induced drag of the system is the same, although the individual birds experience different induced drags: (a) front elevation; (b) V-formation; (c) echelon formation.

In these large flocks the birds are distributed in the fore-and-aft directions and are separated vertically and laterally, in the spanwise direction. Munk's stagger theorem indicates that the fore-and-aft distribution does not affect the overall lift-dependent drag of the formation. If the birds, of wingspan b, project onto a plane normal to the flight directions forming a flock in the shape of an ellipse of height, H, and breadth, B, and if, additionally, the birds are randomly spaced with an average distance between birds denoted by d then the following result is obtained. The decrease in the vortex drag expressed as a ratio to the vortex drag in solo flight is given by

$$\frac{\pi^3}{64}\left(\frac{b}{d}\right)^2\left(\frac{B-H}{B+H}\right)$$

An aerodynamic advantage is seen to accrue only if the breadth of the flock exceeds its height, the disadvantageous influence of height being a factor which we could have anticipated from our earlier discussions.

6.13 VORTEX THEORY OF BIRD FLIGHT

Up to the present stage we have discussed the flapping flight of birds simply by adapting our knowledge of the steady-state aerodynamics of fixed-wing aircraft to the more complex motions under discussion. This has had the considerable advantage of familiarity. However, as flapping flight is by its nature unsteady, to discuss it in terms of quasi-steady aerodynamics, on occasions we have had to accept certain logical difficulties (see section 6.7).

An alternative approach to the interpretation of flapping flight is to represent the flow developed by the wing in a way which is consistent with its inherently unsteady character. This is the basis of a new method (Rayner, 1979), the vortex theory of flapping flight, which is applicable to both hovering and forward flight and explains the aerodynamic properties of the wing in terms of vorticity shed by the wings at the end of each powered stroke.

For the moment let us digress to consider briefly a familiar situation. When an oarsman removes his blade from the water at the end of a powerful stroke he leaves behind a swirling vortex which is clearly visible on the surface of the water. Trailing behind a boat with several crew members there is a regular array of vortices. We can relate this experience to the situation under consideration. In the case of a bird flapping its wings the vorticity created is not visible, but by analogy we deduce that the vorticity shed at the end of each downstroke takes on the shape of a small-cored vortex ring, similar in shape to the smoke rings sometimes seen above a lighted cigarette in still air. In the case of hovering flight the vorticity regularly shed after each wing beat causes a vertical stack of coaxial, circular vortex rings to be formed, whereas in forward flight the rings are elliptical in shape, are more widely separated, and their axes are inclined to the vertical.

Unlike the adaptations of steady-state aerodynamics, the vortex theory does not involve the use of lift or drag coefficients. Instead, in formulating the force balance for the bird, attention is focused on the rate of increase of momentum in the wake induced by the beating wings, and this must just equal the lifting and thrusting forces required to maintain flight. The rate of working by the bird's wings to generate the induced flow is equal to the rate of increase of kinetic energy in the wake, and this is called the induced power.

In hovering flight the wake is used for weight support only, and the distribution of vorticity in the wake depends upon two non-dimensional parameters: the

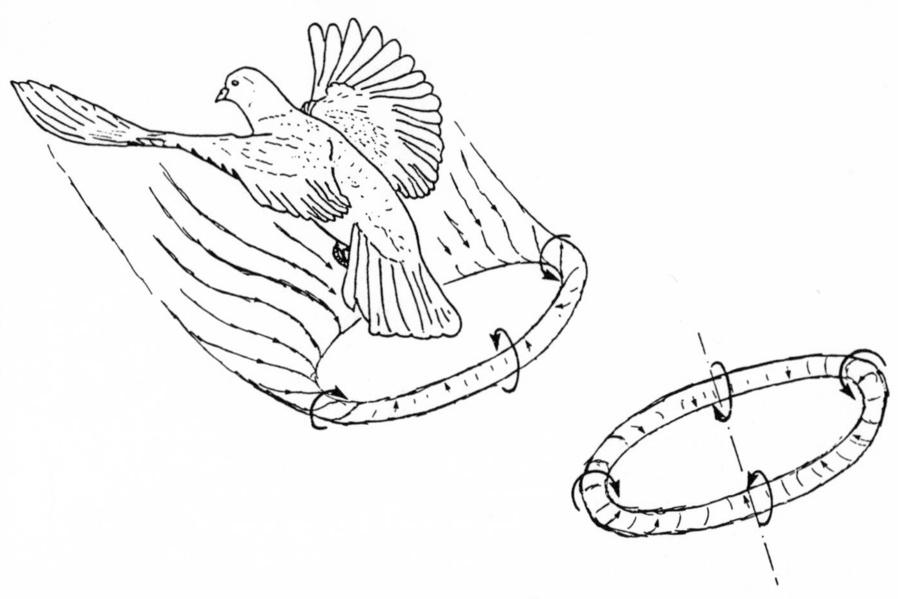

Figure 6.10. Schematic representation of the creation of discrete vortex rings during flapping flight. Initially the trailing vorticity shed by the wing consists of a complex-shaped surface, which rolls up to form a vortex ring for each beat of the wing. In forward flight the rings are elliptical with the major axis aligned with the direction of flight (projection effects mask this property in the diagram). The rings forming a vortex chain are displaced relative to each other in both the horizontal and vertical directions and are neither horizontal nor coaxial. In hovering flight the wake consists of a vertical stack of horizontal, coaxial, circular vortex rings.

feathering parameter, f, and the non-dimensional vortex ring radius, R, which typically has a value just less than 1.

The feathering parameter is defined by

$$f = \frac{mgT_w^2}{2\pi^3\rho s^4} \tag{6.28}$$

where m is the mass of the bird, g is the acceleration due to gravity, T_w is the period between successive wing beats, ρ is the density of the air and s is the wing semi-span, $b/2$.

An insight into the physical interpretation of f may be gained in the following way.

Using momentum jet theory, the mean induced downwash velocity of the air at the plane of the wing, $u/2$, is given by an analysis similar to that leading to equation (6.5). Thus

$$\frac{u}{2} = \left(\frac{mg}{2\rho A_D}\right)^{1/2} = \left(\frac{mg}{2\pi\rho s^2}\right)^{1/2} \tag{6.29}$$

The quantity $(mg/2\pi\rho s^2)^{1/2}$ is therefore a measure of the actual downwash velocity induced by the motions of the wing.

A corresponding measure of the average velocity of the wing itself is provided by the mean wingtip velocity, u_t, defined by

$$u_t = \frac{\pi s}{T_w} \tag{6.30}$$

Substitution of equations (6.29) and (6.30) in equation (6.28) yields

$$f = \left(\frac{u}{2u_t}\right)^2 \tag{6.31}$$

showing that $f^{1/2}$ is a measure of the ratio of the mean induced velocity of the air in the vicinity of the wing to the average velocity with which the wing tip moves.

For small values of f the wake predicted by vortex theory is similar to the jet wake of actuator disc theory. With increasing f the spacing between successive vortex rings increases and the flow fields predicted by the separate theories diverge, that of steady-state theory becoming increasingly unrealistic. For efficient hovering the vortex theory indicates that the strategy a bird should employ is to increase both wingbeat frequency and stroke amplitude as far as is consistent with body structure.

In horizontal forward flight the momentum of each vortex ring must balance the profile drag of the wing, the profile drag of the body (formerly referred to as parasite drag), and the bird's weight. The power required for sustained flight is the sum of the induced power and the profile power, the latter being the power needed to overcome the profile drags of the wing and body.

At low flight speeds the induced power is dominant whereas the profile power predominates at high flight speeds. The variation of power required with flight speed allows two characteristic flight speeds to be defined, at which the work done per unit time and per unit distance are at a minimum. These velocities correspond to the velocities V_{mp} and V_{md}, respectively, derived from steady-state aerodynamics.

The above results for forward flapping flight are similar to results we had already deduced by applying steady-state aerodynamic theory. Where the vortex theory provides a real advance is in highlighting the relationship between the kinematics of wing movement and efficient flight. Birds with low values of the feathering parameter, i.e. birds with high wingbeat frequency and/or large wing span, are best able to hover and to fly forward slowly. A bird gains advantages in reducing power consumption if it is able to change its wingbeat frequency and stroke amplitude to suit particular flight conditions. Thus in slow forward flight a large-amplitude, rapid downstroke is needed, whereas in fast forward flight a small-amplitude stroke reduces power consumption.

The application of vortex theory to analyse bird flight is in its infancy. It is to be expected that it will yield many fruitful results when further put to the test.

Chapter 7

Insect Flight

Throughout the world entomologists have identified between 750,000 and one million different species of insect, and thousands more are discovered each year. They comprise some three-quarters of the total of animal species which have been discovered, named and described.

Their abundance can be ascribed to three main factors—adaptability, size, and their ability to fly. Insects have exploited an extraordinary range of ecological niches. They are to be found from the polar regions to the equators, in such disparate environments as the snow-covered slopes of mountains, dark caves, salt lakes, hot springs, and barren deserts (Imms, 1971).

Insects display an astonishing diversity of feeding habits. Some are parasitic; every kind of flowering plant is used for food by at least one species of insect; decomposing organic matter supports many others. In their struggle for survival their small size has proved an advantage. A minute quantity of food, sufficient to nourish an insect, would be quite inadequate for a larger animal. They are able to find shade, shelter, and refuge in places inaccessible to other creatures. Flight has conferred on insects the freedom to escape from predators, an ability to gain ready access to food supplies, and the mobility to more easily find a mate.

The main flight styles utilized by insects are hovering and slow forward flight, though some are capable of gliding and others have demonstrated an ability to fly forward at relatively high speeds.

7.1 ANATOMY

The body of an adult insect consists of three distinct parts. The head carries the mouthparts and sensory organs, such as the eyes and antennae, and is principally concerned with feeding and decision-making. The thorax is the seat of the insect's means of locomotion; it contains the necessary musculature, and attached to it are three pairs of legs and, usually, one or two pairs of wings. The abdomen contains the organs concerned with digestion, breathing, excretion, and sexual activity.

Insects have an external skeletal structure. Roughly in the shape of a cylinder it surrounds the body, and not only is it a highly efficient form of construction, providing the greatest flexural and torsional strength for a given amount of material, but it also affords an excellent means of protection for the internal organs against dehydration and penetrative damage.

Unlike the wings of birds, which contain their own muscular system, insects' wings possess no internal muscles but are simply aerodynamic surfaces powered from within the thorax. In cross-section, an insect's wing profile is quite different from that of an aircraft or bird. It is not smooth; instead the presence of veins causes the surface to display distinct ridges. The veins are thicker and closer together towards the leading edge of the wing, the actual pattern of the veins varying from one species to another, thereby constituting one of the principal means at the disposal of the entomologist in identifying and classifying different species of insects.

Inspection under the microscope reveals that an insect wing does not usually consist simply of veins and cells; amongst other features revealed by detailed examination are tiny hairs or scales, which may cover much of the wing membrane; in some species there are concentrations of bristles on the leading edge veins and in others the trailing edge may carry a fringe of hairs.

The brightly coloured scales carried by butterflies and moths play an important role in attracting a sexual partner and, in some cases, the coloration is also a warning sign to potential predators. It is not so widely known that these scales also have an important influence on the aerodynamic qualities of the wing; wind-tunnel tests have shown that the lift generated by a moth's wing from which the scales have been removed is significantly lower than that generated by the basic wing at the same angle of incidence.

The earliest flying insects—which appeared during the late Carboniferous period, some 270 million years ago—had two pairs of wings which functioned independently, and this system has survived to the present day in certain insects such as the dragonfly (Odonata), cockroaches and grasshoppers (Orthoptera), and mayflies (Ephemeroptera). However, this system is aerodynamically inefficient and many insect orders have demonstrated an evolutionary convergence on more efficient arrangements. Insects such as bees and wasps (Hymenoptera) and butterflies and moths (Lepidoptera) which have retained two pairs of flapping wings have evolved mechanisms which interlock the forewing and hindwing together so that they work in unison. In other insect orders, only one of the pair of wings is involved in flapping motion, the other pair being held in a fixed position in flight. The beetles (Coleoptera) have followed this evolutionary tendency, the hindwings playing an active role in flight, whilst the forewings have become transformed to perform a dual role, acting as protective covers, called elytra, when the insect has alighted, but being spread to provide a fixed aerodynamic lifting surface when the insect is in flight. In some insect orders the hindwings have disappeared entirely, leaving a single pair of flying surfaces.

This evolutionary process has taken place within the order of the true flies (Diptera).

The construction of the thorax plays an extremely important role in the flight of insects. Within it are contained the muscles which are responsible for the wing movement, and their function will be discussed shortly. The external skeleton consists of a hard cuticle, with good elastic properties, and this encloses a volume of a high-quality substance, not unlike vulcanized rubber. This substance, resilin, responds elastically under load and exhibits little internal friction. The thorax is capable of appreciable deformation under the action of an applied force.

Amongst the dipteran flies the geometry of the thorax, together with the wing pivots, is such that the only stable positions of the wings occur at the extremes of movement. In this respect the mechanism resembles an electric light switch; perhaps an even closer parallel to the role of the thorax is provided by the old type of biscuit tin, the walls of which could be clicked between extreme positions.

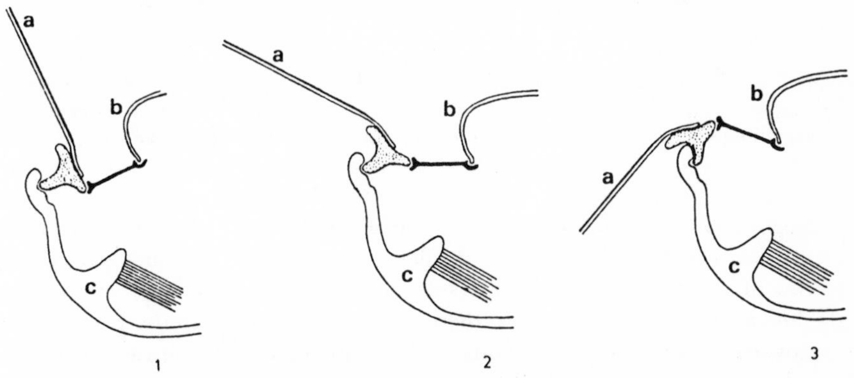

Figure 7.1. Schematic representation of the click mechanism. (1) Shows the wing held in the stable up position; (3) the wing is in the stable down position. The intermediate unstable position is shown in (2): (a) wing; (b) thorax roof; (c) thorax wall. (Reproduced, with permission from Nachtigall, W, (1974), *Insects in Flight*. Allen & Unwin, London.)

Once set into motion, the wing–thorax system is capable of continuous oscillation at a natural frequency determined by the elastic properties of the thorax and the inertia of the oscillating mass; the only additional energy that has to be supplied is required to overcome the external work done by the wings and any slight dissipation due to friction within an extremely efficient elastic system.

It is worth highlighting the superb efficiency of the flight system of insects. Every time a bird beats its wings the muscles are required to provide sufficient effort to accelerate and decelerate the wings. Put another way: the muscles have to overcome the inertia of the wings. With insects the elastic material of the thorax makes this particular form of muscular effort unnecessary by constantly interchanging the kinetic energy of the wing motion with strain energy stored in the deformed thorax.

In the lower orders of insects, such as the grasshoppers and crickets (Orthoptera) and dragonflies and damselflies (Odonata), movement of the wings

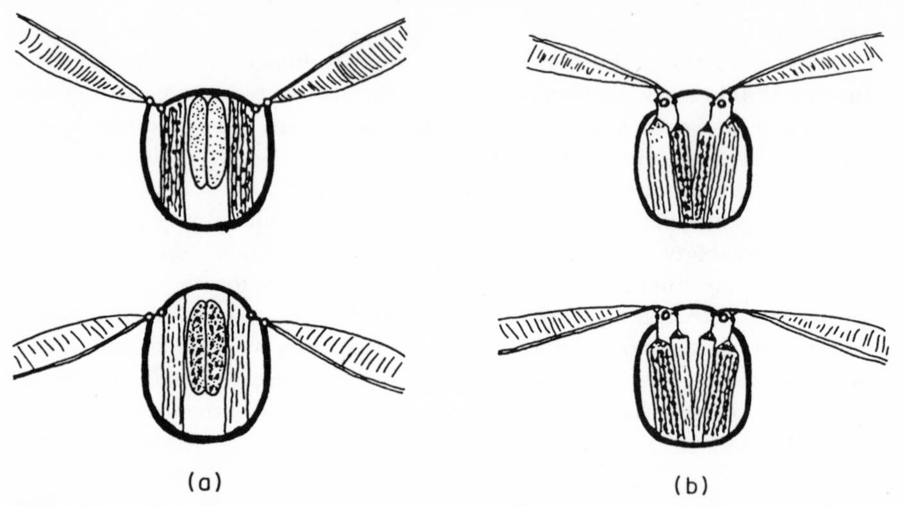

Figure 7.2. Flight musculature. Contracting muscles are shown stippled: (a) indirect flight muscles, typical of flies and midges; (b) direct flight muscles, typical of dragonflies and grasshoppers. In each case the upper diagram illustrates the upstroke and the lower diagram shows the downstroke.

is brought about in the following way. Attached to the underside of each wing are two muscles which are anchored to the floor of the thorax. The wing is hinged to the top of the thorax on its upperside, and on the underside, between the attachment points of the muscles, there is a point about which the wing pivots. Downward movement of the wing is brought about by contraction of the outermost muscle; upward movement results from contraction of the inner muscle. Each of the four wings therefore has its own independent power supply, and this accounts for the fact that the movement of the pairs of wings is often uncoordinated. The maximum wingbeat frequency of which insects with this system are capable is between about 25 and 100 cycles per second, a figure which is determined by the fact that the muscles they possess need to be activated once per cycle by nervous impulses.

Movement of the wings in the advanced insects is brought about by a more sophisticated system, with two separate sets of muscles—the indirect and the direct muscles, respectively. The indirect muscles are the largest in the body and their name reflects the fact that both ends of the muscles are connected to the walls of the thorax and not to the wing itself. Movement of the wings in the vertical direction is the responsibility of these indirect muscles, of which there are two pairs—the vertical and the longitudinal—which contract and relax alternately. Contraction of the vertical indirect muscles causes a depression in the roof of the thorax and, because of the manner in which the wings are hinged to the thorax, this movement results in the wings being raised. Contraction of the longitudinal muscles causes the roof of the thorax to curve upwards, and as a consequence the wings move downwards. The direct wing muscles are attached to the roots of the wings and they have two principal functions: to move the wing in the fore-and-aft direction and also to rotate the wing about its spanwise axis.

Many of the advanced insects are powered by fibrillar muscles which are capable of sustained oscillation, without the need for each contraction to be triggered by a separate nerve impulse. With this musculature tuned to the elastic properties of the thorax, very high wingbeat frequencies are possible and many bees, flies, and wasps are capable of beating their wings at 250 cycles a second, and some members of the Diptera family attain wingbeat frequencies of over 1000 cycles per second.

The distinction between insects whose wing movements are brought about by muscles of the fibrillar type and those with flight muscles of the neurogenic type is most evident when the insect loses part of its wing through damage. This occurrence causes an increase in wingbeat frequency to result in insects powered by fibrillar muscles, because the reduced moment of inertia and mass of the wing lead to an increase in the natural frequency of the oscillating mechanical system. On the other hand, insects with muscles of the neurogenic type depend upon a nerve impulse via the central nervous system to initiate muscle contraction, and this process is unaffected by the damage to the wing. Consequently in this type of insect the wingbeat frequency remains unchanged after damage.

The factor which ultimately determines the maximum size of insects is the method of breathing employed. They inhale and exhale through a set of orifices called spiracles, placed on each side of the abdomen. Insects have no lungs, but instead use a complex set of air-tubes, or tracheae, which service the internal organs of the body. The air-tubes form an elaborate system, constantly branching, becoming smaller and smaller in diameter in the process, and ending in the form of capillaries of microscopic proportions, the tiniest being less than 1 micrometre (10^{-6} m or 4×10^{-5} in.) in diameter. In these fine tubes, oxygen and carbon dioxide are conveyed by diffusion, a process which is effective only over very short distances. Many small insects rely solely upon diffusion for transport throughout the entire tracheal system, and it is this factor which governs the maximum dimensions of the bodies of such insects. For larger insects the air-tubes are longer and if these insects were to rely upon diffusion alone, they would experience oxygen starvation and would be unable to survive. So, when the thorax and abdomen are more than a few millimetres in diameter, the primary tracheae must be ventilated. Larger insects principally employ one or other of two mechanisms for this purpose. Some insects use abdominal movements which result in abdominal pumping of the air, whilst others, including the large dragonflies, exploit the movement of the wings to bring about movements of the thoracic walls, which leads to thoracic pumping at the wingbeat frequency. In some large beetles yet another mechanism is used for ventilation in flight. Forward movement of the body through the air is responsible for variations in pressure being established over the external surface of the insect; differential pressures are exploited for the direct ventilation of one or two pairs of giant primary air trunks.

7.2 THE AERODYNAMICS OF INSECT FLIGHT

Insect wings function at modest Reynolds numbers based on mean flight speed. For example, locusts (*Schistocerca*), with a mean wing chord of 2 cm, are amongst the larger insects. They fly at forward speeds of about 3.5 m s^{-1} and so, taking the kinematic viscosity of air at sea level as 1.5×10^{-5} m^2 s^{-1}, the flight Reynolds number is computed as $3.5 \times 0.02/1.5 \times 10^{-5} = 4700$, which is about 4 per cent of a pigeon's Reynolds number. At the opposite extreme there are tiny insects which have a Reynolds number of the order of 10^{-3}, and under such conditions these insects resort to flight styles quite different from their larger relatives.

At present our understanding of the detailed motion of the air about the wings of insects is fragmentary; in the main, current knowledge of insect flight is largely concerned with the overall aerodynamic performance of insects as flying machines.

A number of insects insert an occasional gliding phase into a programme of horizontal flight. Generally it is the larger insects which are the most accomplished gliders and amongst their number are the swallowtail butterfly (*Iphiclides podalirius*) and several species of the dragonfly family (Aeschna). By and large, though, it is a flight style which most insects cannot, or fail to, exploit.

Beating of the wings is therefore almost always present during insect flight. The earlier discussion of the physiology of the thorax and its musculature has already indicated that insects are capable of very high wingbeat frequencies. Some data for a range of insects are contained in Table 7.1.

Although the wing of an insect deflects and twists due to the aerodynamic loads it sustains, once an insect has unfurled its wings from their resting positions the planform it adopts is constant and predetermined. The geometry of insect wings cannot be adjusted in the way that birds modify the shape of their wings in flight, and so this method of controlling flight performance is not available to insects. Instead, flight control is effected largely by the choice of the mean stroke plane about which the wings move.

First of all we consider insects with a single pair of wings or coupled wings which work in unison. When such an insect traverses across country by means of fast forward flight, it does so using wing movements very similar to those employed by a bird engaged in the same process. Thus the mean stroke plane of the wings is set at an acute angle relative to the vertical and, in relation to the body of the insect, the wings describe a narrow figure-of-eight about the mean stroke plane. In this way components of thrust and lift are generated to balance the aerodynamic drag and body weight respectively. For the purpose of hovering, the body is held upright and the wings beat about a mean stroke plane which is essentially horizontal, and again the path of the wing tip is a narrow figure-of-eight. At flight speeds between those for hovering and fast forward flight the mean stroke plane assumes an intermediate angle between the horizontal and vertical, the wing tip path maintaining its figure-of-eight form.

There are exceptions to the above general description of hovering flight.

Table 7.1 Wingbeat frequency of some insects.

Insect	Wingbeat frequency (Wingbeats per second)
Cabbage white butterfly	12
Damselfly	16
Desert locust	20
Dragonfly	35
Cockchafer	45
Cranefly	55
Hawk moth	80
Common wasp	100
Horsefly	100
Hornet	100
Bumblebee	150
Deer botfly	160
Hoverfly	180
Honeybee	200
Housefly	200
Blowfly, bluebottle fly	200
Mosquito	300
Midge (Forcipomyia)*	1045

* Fastest recorded.

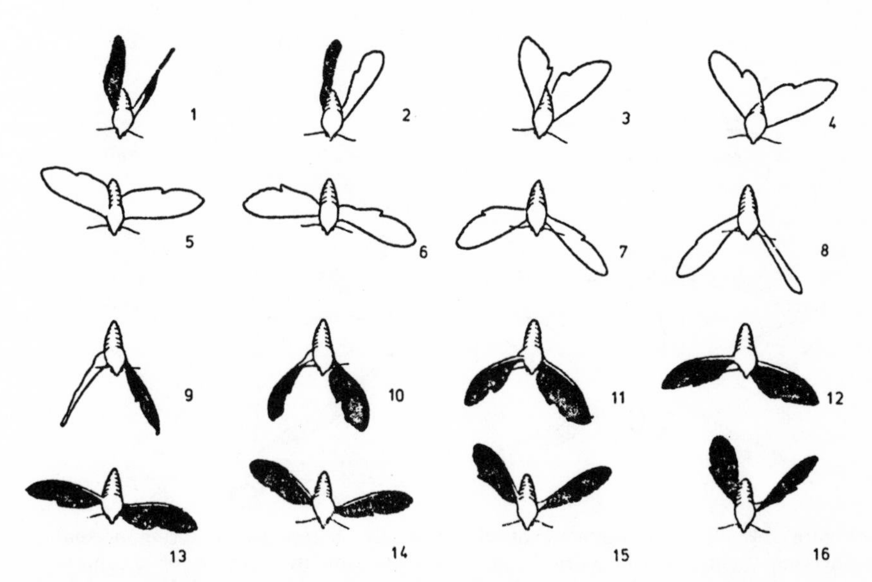

Figure 7.3. Wing movements of the moth *Manduca sexta* during normal hovering flight. The underside of the wings is shaded black to assist in the interpretation of the complex movements. The camera was mounted vertically above the hovering moth. (Reproduced, by permission of the Company of Biologists, from Weis-Fogh, T. (1973), *J. Exp. Biol.*, **59**, 169–230.)

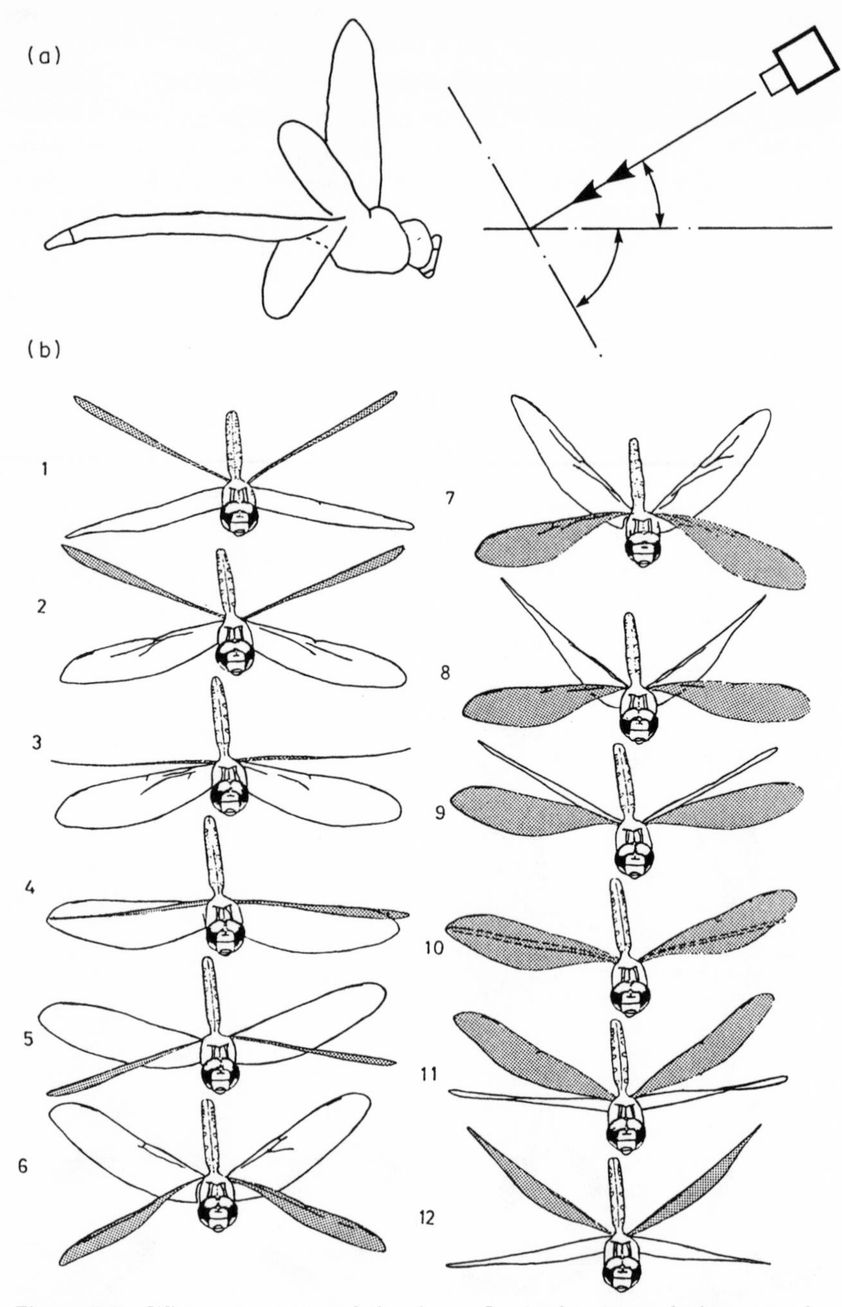

Figure 7.4. Wing movements of the dragonfly *Aeschna juncea* during normal hovering flight: (a) side view of the insect showing the position of a camera mounted normal to the stroke plane; (b) successive positions of the wings as viewed by the camera—the forewings are shaded. (Reproduced, with permission from Norberg, R. (1974). In Wu, T. *et al.* (eds), *Swimming and Flying in Nature*, vol. 2. Plenum Press, New York.)

Detailed observations (Norberg, 1975) of the dragonfly *Aeschna juncea*, which has two pairs of wings which operate independently, have revealed that the wing movements are quite different from those of other insects (and the humming-birds) in hovering flight. Whilst the body is held almost horizontal, both wings move more or less identically with the wing stroke plane maintained at an angle of about 60° to the horizontal, the movement of the wing tip being confined to this single plane during the downstroke and upstroke.

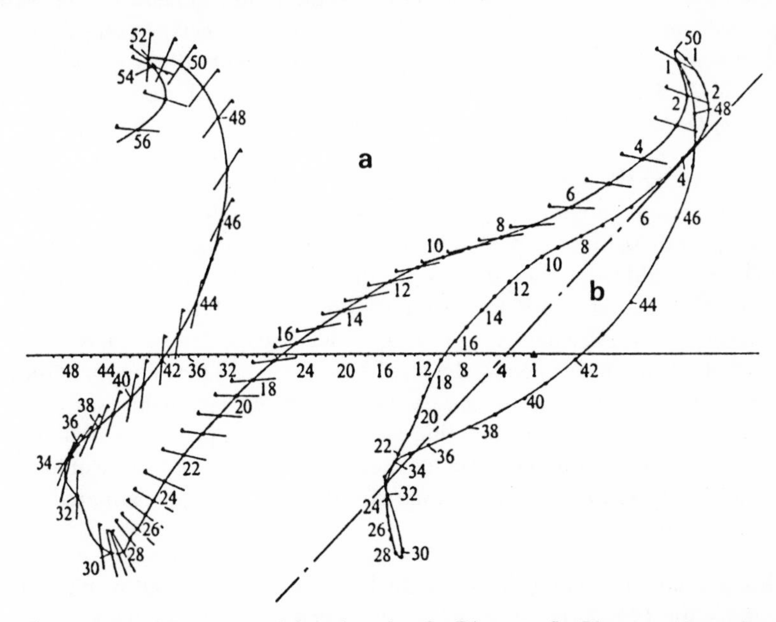

Figure 7.5. Movement of the wing tip of a Dipteran fly *Phormia regina* as it flies from right to left. A triangular symbol is used to denote the upper leading edge of the wing, which is denoted by a short line. The open loop (a) shows the movement of the wing through space. During the upstroke the wing actually moves backwards for a short period each stroke. The closed loop (b) shows the movement of the wing relative to a point fixed in the insect. For example, the insect would see its own wing movement as depicted in (b). Each unit on the horizontal axis indicates the distance by which the head or thorax of the fly advances between one position of the wing and the next. (Reproduced, by permission of the publishers, from W. Nachtigall, (1974). *Insects in Flight*. Allen & Unwin, London.)

In addition to the basic figure-of-eight movements of the wing of insects during hovering and forward flight, large amounts of twist are also exhibited. During the downstroke the wing is approximately horizontal; during the upstroke the leading edge is sharply pitched up and in extreme cases, as the diagram (Fig. 7.5) depicting the motion of the wing of the fly *Phormia regina* shows, the wing may even be rotated so that the leading edge points in the downstream direction during part of the wingbeat cycle. Rapid rotation of the wing is mainly concentrated in that part of the wingbeat cycle near the top and bottom of the stroke.

7.3 THE FLIGHT SPEED OF INSECTS

Insects do not fly for long at constant speed and, because their airspeeds are comparatively low, even a slight breeze is sufficient to substantially influence the speed of an insect over the ground. For many years these and other factors—such as the difficulty of persuading the creature to fly between a specified starting line and finishing tape—so affected the measurement of the flight speed of insects that few reliable data were available. Indeed some quite remarkable claims have appeared concerning the flight capabilities of certain insects. As recently as 1927 the following statement concerning the flight of the deer botfly (*Cephenemyia pratti*) was published by Dr Charles Townsend as a serious contribution to science in the *Journal of the New York Entomological Society*:

> Regarding the speeds of Cephenemyia, the idea of a fly overtaking a bullet is a painful mental pill to swallow, as a friend has quaintly written me, yet these flies can probably do that to an old-fashioned musket ball. They could probably have kept up with shells that the German big-bertha shot into Paris during the world war. The males are faster than the females, since they must overtake the latter for coition.... On 12,000 ft summits in New Mexico, I have seen pass me at an incredible velocity what were quite certainly the males of Cephenemyia. I could barely distinguish that something had passed—only a brownish blur in the air of about the right size for these flies, and without sense of form. As closely as I can estimate, their speed must have approximated 400 yards per second.

If this estimate of the flight speed of the botfly were accurate, it would in fact be moving some 10 per cent in excess of the speed of sound! Even so, for the next 10–12 years after this article had been published the reliability of this figure was widely accepted and the botfly was regarded as the fastest creature on the face of the earth.

In an editorial, the *New York Times*, commenting upon a new airspeed record of just over 300 miles per hour established by a seaplane, warned its readers not to be too boastful of mankind's achievements, since the deerfly has a speed of 700 miles an hour. Some years later, in 1938, the *Illustrated London News* contained a two-page feature quoting comparative speeds of man, animals, insects, etc. A speed of 818 miles per hour was given for the male deer botfly, and on the next page there was a photograph of this insect captioned, 'The speed champion of the world, capable of outstripping sound'.

However, some scientists became sceptical about these extraordinary figures and in 1938 Langmuir evaluated the implications of such claims. He showed, by considering the energy expended in overcoming drag and the pressure sustained on the head of the insect, that the claimed flight speed was quite unrealistic.

As a result of Langmuir's work a figure of 25 mph (approximately 11 m s^{-1}) is now widely accepted as the maximum speed of a botfly, and is generally regarded as the highest flight speed which any insect attains. Yet even this figure deserves closer inspection. Langmuir made a simple model of the botfly using a lump of solder of the appropriate size, which he rotated on the end of a piece of thread. He concluded that at a speed of about 25 mph the visual impression best corresponded to the 'barely distinguishable blur' reported by Townsend. Langmuir calculated that under these conditions the botfly needed to consume about 5 per cent of its body weight in food every hour. The improved estimate of flight speed proposed by Langmuir ultimately rests upon the interpretation of visual impressions.

It is important to recognize that, even if this process led to an accurate estimate of flight speed—a conclusion which itself is debatable—this speed must be interpreted as a groundspeed, and not an airspeed.

Air currents are a feature of the environment near the summit of hills and mountains, and it is just when an insect derives some benefit from a tail wind that its speed is most likely to impress itself on the observer. A modest tail wind of, say 3 m s^{-1}, is sufficient to boost an airspeed of 8 m s^{-1} to an observed groundspeed of 11 m s^{-1}. Also at an altitude of 12,000 ft atmospheric density is approximately 70 per cent of the sea level value. On account of this decrease in density the maximum flight speed will be greater at altitude than at sea level.

Estimates of the required and available power can be made in the following way. Taking the mass of the insect as 0.2 g, the maximum available power output of the flight muscles as 220 W kg^{-1}, and assuming that 25 per cent of the insect's body weight is made up of flight muscles, the available power is computed as 0.011 Watts. An approximate figure for the drag area of the insect's body might be $SC_D = 10^{-5}$ m^2. To overcome this drag at an airspeed of 11 m s^{-1}, power would need to be expended at a rate of 0.008 Watts at sea level or 0.0056 Watts at 12,000 ft. Additional power would be required to overcome the corresponding drag on the insect's wing, and the power available for flight would be diminished by the mechanical efficiency of the muscular system. It is not possible to quantify these additional factors with any great precision, but it is already evident that an energy balance consistent with sustained flight at an airspeed of 11 m s^{-1} is improbable. As the required power increases as V^3, a maximum airspeed nearer 8 or 9 m s^{-1} would seem a more reasonable estimate at sea level.

Until some accurate measurements are taken in the field, including the wind speed as well as the groundspeed of the insect, the question of the maximum speed of the botfly will remain open.

The airspeed which a range of insects can achieve in sustained flight at sea level is given in Table 7.2.

Compared to the flight speeds of birds, the performance of insects at first appears extremely modest. However if the data, instead of being quoted in terms of distance covered in unit time, are expressed in terms of body lengths covered in unit time, a very different picture emerges. Typical figures for birds are: swan 10 body lengths per second, swift 60. In contrast a fly may cover as much as 250–300 times its own body length each second. By way of comparison, in a sprint race a man covers about 5 body lengths each second.

Table 7.2 Typical flight speeds of some insects.

Insect	Typical flight speed	
	mph	$m\ s^{-1}$
Mayfly	1.1	0.5
Mosquito	2	0.9
Stag beetle	3.4	1.5
Damselfly	3.4	1.5
Housefly	4	1.8
Cockchafer	5.6	2.5
Cabbage white butterfly	5.6	2.5
Common wasp	5.6	2.5
Bumblebee	7	3
Blowfly, bluebottle fly	7	3
Honeybee	7	3
Hoverfly	8	3.5
Desert locust	10	4.4
Hawk moth	11	5
Horsefly	14	6
Hornet	15.5	7
Dragonfly	17	7.5
Deer botfly* (*Cephenemyia pratti*)	20	9

* Probably the fastest insect

Some specialized forms of insect are capable of living at extremely high altitudes, far above the timber-line, in spite of the difficulties associated with the reduction in temperature, oxygen levels, and air density, all of which impair activity in general, and aerial locomotion in particular (Mani, 1962). In the Himalayas butterflies (Lepidoptera) of the species *Parnassius acco* occur up to 5800 m (19,000 ft), and in this region there are various species of true flies (Diptera) found at 6000 m (19700 ft), which is the highest altitude for any flying form of insect anywhere in the world.

7.4 THE FORWARD FLIGHT OF LARGE INSECTS

Insects such as damselflies and dragonflies (Odonata), grasshoppers and locusts (Orthoptera), butterflies and moths (Lepidoptera), and bees and wasps (Hymenoptera) all exploit aerodynamic principles similar to those utilized by birds for the purposes of achieving sustained flight. The rapid movement of the wings generates a wake in which the rate of increase in momentum is sufficient to balance the combined forces due to the weight of the insect and the profile drags of the wings and of the body of the insect. This flight style is utilized by all large insects because, at the Reynolds numbers at which they fly, the inertial forces they experience are much larger than the viscous forces and it is therefore appropriate to use these inertial forces as the prime means of securing aerodynamic lift.

7.5 THE FLIGHT OF TINY INSECTS

With the correlation that exists between the size of insects and their flight speed—broadly speaking the largest insects fly fastest and the smallest fly slowest—it follows, due to both the reduction in size and flight velocity, that there is a very rapid reduction in Reynolds number with size under the conditions at which tiny insects operate. For the tiniest insects, moving at Reynolds numbers below 1, the inertial forces they experience are over-shadowed by the viscous forces arising from their motion through the air. In these circumstances no longer can an insect efficiently exploit inertial forces. Instead it is necessary for the insect to use a radically different means of locomotion through the air, by turning the viscous properties of the air to advantage.

Amongst the tiniest of flying insects are the thrips (Thysanoptera), of which there are over 5000 species varying in length from about 1 mm (0.04 in.) to 14 mm (0.55 in.). The wings of four different members of this order are shown in Fig. 7.6. They are quite unlike the wings of larger insects. All four are very similar to each other, consisting of a main stem which carries a multitude of regularly arrayed bristles, all set in the same plane.

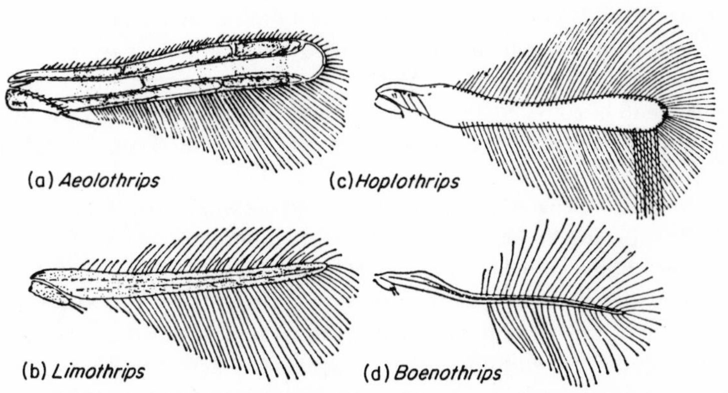

Figure 7.6. Some examples showing the diversity of wing geometry of various species of thrips (Thysanoptera). The wings of these tiny insects are less than 1 mm in length. (Reproduced, with permission from Lewis, T. (1973), *Thrips*. © Academic Press Inc., London.)

Whereas the membranous wings of the larger insects form a continuous surface impermeable to the air, the hairy wings of thrips are full of gaps through which the air is free to pass.

The form of the wings of thrips is the result of evolutionary convergence to exploit the viscous world in which they find themselves. To the thrip the air is as sticky as we find treacle. As a consequence of the Reynolds number effect it is a world dominated by viscous drag; aerodynamic lift is unavailable.

At present it is not known for certain just how the wings of a thrip function. One hypothesis that has been advanced (Kuethe, 1975) is that the beating of the wings causes waves to propagate in the chord-wise and span-wise directions amongst the bristles. It is surmised that the mechanism by which a propulsive force is thereby generated is similar to that of the rotating tails (flagella) of spermatozoa, which also travel at very low Reynolds numbers.

Thrips are not alone in their approach to movement through the air. Tiny members of the families of beetle (Coleoptera) and bees and wasps (Hymenoptera) also possess the same hairy wings. Like the thrips, the small Hymenoptera have two pairs of wings, whereas the small beetles only use one pair of wings for flight, since the forewings function as wing cases.

7.6 SWARMING FLIGHT

In some species of insect, large numbers take to the wing simultaneously. We consider two distinct examples.

7.6.1 Mayflies

Certain fish, such as the trout, can be persuaded to rise to artificially constructed flies. In this context of special interest to anglers are the Ephemeroptera, the mayflies. These insects are also a worthy subject for the student of aerodynamics. The mature adult is known as a spinner. As a prelude to mating, male spinners collect together on the wing to form a swarm, and in larger swarms up to several thousand insects may be present. Within the swarm the individual insects hold an approximately constant position in space, and the boundary of the swarm itself assumes a well-defined position relative to the water's edge. A light breeze, rather than still air, provides the optimum flight conditions. Basically the spinners face into the wind and fly forward into the wind at the same speed as the wind blows the insect backwards, although different species of mayfly play different variations on this basic theme (Harris, 1956).

Some species are capable of maintaining a virtually fixed position in space, adopting a flight style exactly the same as the avian hovering of the kestrel. Other species rise and fall continuously; a brief period of active beating of the wings causes the insect to climb almost vertically and then the wings are held fixed in a pronounced dihedral as the insect glides gently towards earth again with its tails trailing behind (Fig. 7.7). Yet other species intersperse a short period of horizontal flight between rising and falling. Whereas it is normally very difficult to study an insect in flight, owing to the rate at which it moves over the ground, the flight of male mayflies may be investigated with comparative ease. Indeed, the expert entomologist is able to identify the species of mayfly on the wing from the subtle variations in flight characteristics (Harris, 1956).

7.6.2 Locusts

Without doubt the most notorious swarming flights of insects are those of the

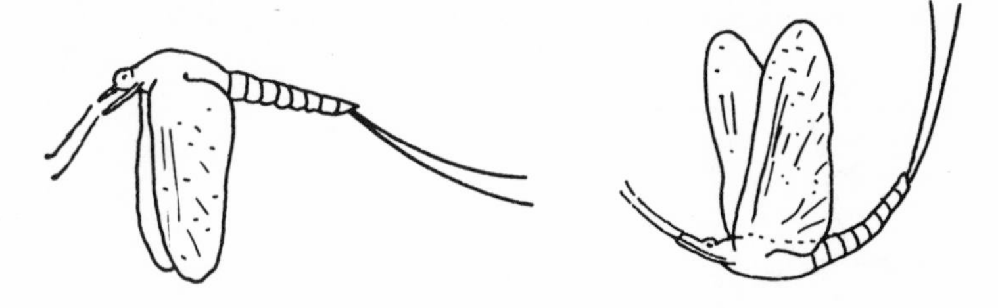

Figure 7.7. The charming flight style of the Ephemeroptera. On the left the mayfly is shown beating its wings and rising. On the right it glides and falls.

desert locusts (*Schistocerca gregaria*), whose habitat is the hot, dry, semi-arid regions of the Middle East, including parts of the African and Asian continents. The Bible refers on a number of occasions to the locust. In the Book of Exodus, Chapter 10, there is the following passage:

> And Moses stretched forth his rod over the land of Egypt, and the Lord brought an east wind upon the land all that day, and all that night; and when it was morning the east wind brought the locusts.
>
> And the locusts went up over all the land of Egypt, and rested in all the coasts of Egypt; very grievous were they; before them there were no such locusts as they, neither after them shall be such.
>
> For they covered the face of the whole earth, so that the land was darkened; and they did eat every herb of the land, and all the fruit of the trees which the hail had left; and there remained not any green thing in the trees, or in the herbs of the field, through all the land of Egypt.

This passage is interesting for a number of reasons: in particular it correctly attributes the gross movement of the swarm to the prevailing wind. Some biologists have believed that these great migrations are as a result of purposeful, directed flight; such is not the case—the swarm is carried along by the motion of the winds (Scorer, 1978). Thus, near a coast, an offshore wind will blow the locusts out to sea; the existence of further food supplies along the coastline will be of no benefit to them.

The way in which the swarm advances is of considerable interest and has been described by Scorer (1978). The greatest concentration of locusts is towards the front of the swarm, where the new food supplies become available. This leading edge rolls forward continuously as locusts from the rear of the swarm advance towards the front. Within the swarm there is a broad circulatory motion (Fig. 7.8). Those insects wishing to gain ground from the rear of the swarm fly at higher altitudes and benefit from increased wind speed with altitude; close to the ground the locusts tend to fly upwind, as this assists the take-off and landing processes by reducing the speed of the insect relative to the ground. At the edges of the swarm, whenever a group of insects flies towards airspace unoccupied by other locusts it turns back towards the main body of the swarm, a mechanism which avoids the

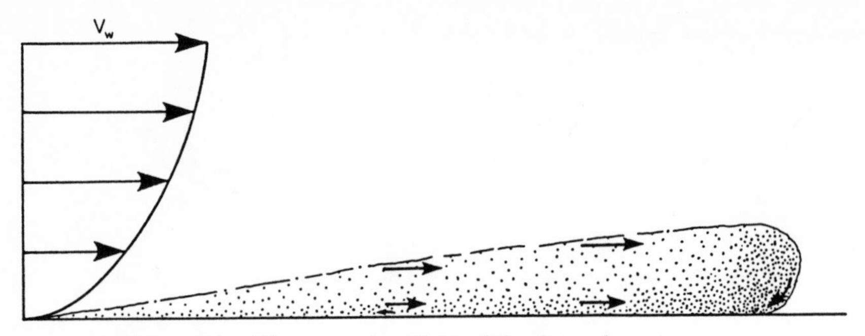

Figure 7.8. The swarming flight of the desert locust.

swarm becoming too dispersed laterally or longitudinally. Thus, at the forward edges those locusts that fly beyond the main concentration of insects turn back before alighting.

Swarms commonly number over a million insects and it was estimated that there were 2.5×10^{11} locusts in a record nineteenth-century swarm. Viewed from within the swarm the motions of the locusts might appear random and haphazard; but, as we have seen, there are broad patterns which govern the movement of the individual insects, and these average out to provide the observed movement of the swarm as a whole.

7.7 THE CLAP–FLING MECHANISM

Until the early 1970s the flight mechanics of insects and of birds in hovering and forward flight had always been discussed in terms of simple adaptations of the steady-state aerodynamics which apply to aircraft flight. By and large this approach seemed to work quite well, but there were occasional difficulties. For example to explain the sustained hovering flight of certain insects extremely high mean lift coefficients were required. An extreme case was that of the small chalcid wasp (*Encarsia formosa*), a small parasitic wasp often used in the biological control of aphids in greenhouses, whose hovering performance implied a mean C_L of about 5. Such a high value was far in excess of the highest values associated with aircraft flight, where maximum lift coefficients rather less than 2 are typical.

It was in 1973 that the distinguished scientist, the late Torkel Weis-Fogh, a Dane who was Professor of Zoology at the University of Cambridge, made a remarkable and inspired contribution (Weis-Fogh, 1973; Lighthill, 1973) to the world of aerodynamics. He showed that the high mean lift coefficients obtained by *Encarsia* were being generated by an aerodynamic mechanism hitherto undiscovered in the world of aeronautics. This mode of aerodynamic lift generation is now known as the clap–fling mechanism. Using high-speed cinematography Professor Weis-Fogh discovered that, in the free hovering flight of *Encarsia*, whilst the movement of the wings during each cycle was in many respects similar to that of other insects, it contained two features of special significance to the generation of lift (Fig. 7.9).

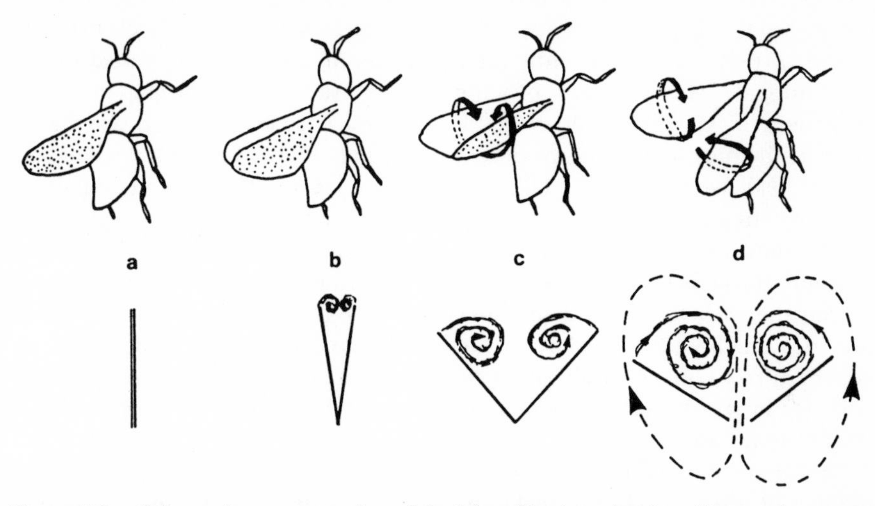

Figure 7.9. Schematic representation of the 'clap–fling' mechanism. Progressive stages are shown from left to right, the lower set of diagrams providing a schematic view of the wing geometry and flow pattern at a typical cross-section of the wing position shown above. (a) The wing surfaces are shown in the 'clap' position. (b) The upper (leading) edges of the wings are parted whilst contact is maintained along the trailing edges. This constitutes the initial movement of the 'fling'. (c) The leading edges are moved progressively further apart and a pair of symmetrical separation vortices are formed between the wing surfaces. (d) The final stage of the 'fling' process. When the angle between the wing surfaces reaches a certain maximum value the trailing edges are finally parted and the wings are moved forward at constant incidence angle, about an axis of rotation passing longitudinally through the body of the insect.

(1) The clap. At the top of the upstroke the wings were brought into contact to form a single vertical surface above the body of the insect.
(2) The fling. Prior to the commencement of the downstroke the leading edges of the opposite wings were flung outward and apart, whilst contact was initially maintained along the trailing edge of the hindwings.

We have seen that the lift generated by an aircraft (or bird) in steady gliding or soaring flight can be explained in terms of the vortex system consisting of the bound vortex surrounding the wing and the pair of trailing vortices which stretch downstream from the wing tips. The maintenance of this vortex system depends upon the continued forward motion of the aircraft. As the aircraft moves through the air, relative to the aircraft the physical properties of the vortex system in the neighbourhood of the wing, such as the geometry of the flow pattern and the strength of the vortex, do not vary with time. This sort of motion is called steady.

With the flapping flight of insects (and birds), relative to the position of the insect's body the positions of the wings are constantly changing with time, and the motion of the air about the wings is inherently unsteady. Professor Weis-Fogh recognized that the clap–fling mechanism was exploited by some insects to create bound vorticity of a significantly higher magnitude than that generated by the conventional steady-state type of aerodynamics. At the end of the clap the wings are essentially stationary relative to the air; as the wing leading edges are moved apart at the start of the fling air moves swiftly in to fill the void in the triangular region between the wings. This induced flow takes the form of a pair of contrarotating vortices of equal and opposite strength.

On completion of the fling the hind margins of the wing separate and the wings continue through their downstroke benefiting from the bound vortex created during the clap–fling. Thus the novel principle of the clap–fling mechanism provides a highly efficient source of bound vorticity for flapping flight, which can either entirely replace, or alternatively can augment, the conventional source of vorticity in steady-state aerodynamics, namely the translation of the aerofoil relative to the air.

Now that the clap–fling mechanism has been observed it is recognized that a number of insects exploit it, including various fruitflies, moths and butterflies, including the common cabbage white butterfly (*Pieris brassicae*) and, though more work needs to be done to provide confirmation, it seems likely that some birds and bats also exploit this mechanism of lift generation. Pigeons, for example, frequently bring their wings together at take-off.

7.8 OTHER METHODS OF LIFT GENERATION

Nevertheless, the discovery of the clap–fling mechanism has not resolved all of the outstanding problems relating to the aerodynamics of animals which generate high lift coefficients. Insects such as some of the hover-flies and dragonflies generate high mean lift coefficients whilst hovering, but careful study has revealed that they do not bring the wings together as in the clap–fling mechanism. Some other aerodynamic mechanism is required to explain the extraordinary flight performances of these creatures.

The insect's wing-tip follows a figure-of-eight path and at the end of the downstroke the wing is flipped over so that the forward edge of the wing is rapidly pitched up. This flip action serves to cast off the bound vorticity from the wing and to prepare for the creation of vorticity of the opposite sense during the upstroke. In contrast to birds, which generate almost all the lift (and thrust in forward flight) during the downstroke whilst the upstroke is used in an essentially passive role, insects generate almost equal amounts of lift during the down- and upstrokes.

The elucidation of the mechanism by which extraordinarily high lift coefficients are achieved is one of the current challenges to aerodynamicists, and it represents a fruitful field of research at the present time. The mechanism of lift generation in fixed-wing aircraft in steady flight has been successfully described

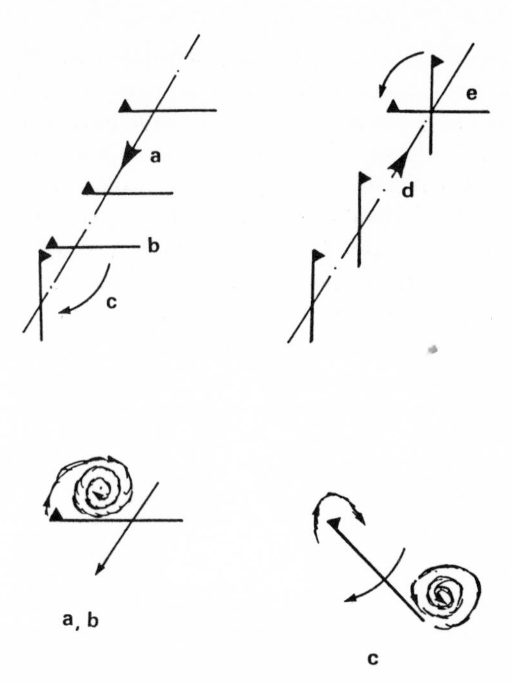

Figure 7.10. Principal features of the wing movements of insects which do not use the 'clap–fling' mechanism. The triangular symbol denotes the upper leading edge of the wing. The mean stroke plane is inclined to the horizontal during forward or hovering flight. (a) Downstroke; (b) pause; (c) supination; (d) upstroke; (e) pronation. Underneath are shown sketches of the main flow patterns contributing to the aerodynamic forces generated during each wingbeat cycle. During the downstroke and pause phases a leading-edge vortex is mainly responsible for the generation of lift. During supination there is a substantial lift component associated with a leading-edge suction force. A trailing-edge vortex which is created at this stage is subsequently shed when the wing is raised during the upstroke.

in terms of a vortex system comprising a starting vortex, a bound vortex, and two trailing vortices, the circulation in the bound vortex system being determined by the Kutta–Joukowski criterion. To explain the high lift coefficients of the animal world which are a feature of hovering flight and low-speed forward flight, it now seems clear that the aerodynamics must be described in entirely different terms, for example in terms of unsteady flow patterns, embodying discrete vortices with high intensities, which are formed and shed with each beat of the wing.

Thus, it has been shown recently that the performance of the dragonfly wing can be explained by focusing attention on two distinct phases of the wing movement in each cycle when the major contributions to lift are generated (Savage *et al.*, 1979). In the first phase, at the start of the downstroke, a rapidly rotating vortex is created and this disposes itself above the wing during the downstroke, thereby creating substantial suction pressures. In the second phase, which occurs in the early stages of the upstroke, the leading edge of the wing is rapidly supinated (pitched up) and this creates a leading edge suction force which, together with a vortex in the vicinity of the trailing edge, leads to a significant augmentation to the aerodynamic lift.

Chapter 8

The Flight of Bats, Pterosaurs, and Other Vertebrates

Bats are the only mammals capable of sustained flight. Several other species of mammal possess anatomical features allowing efficient gliding flight to be performed. Other animals, including the so-called flying fishes, also exploit this flight style, as did the now-extinct pterosaurs. These creatures are the subject of the present chapter.

8.1 BATS

Bats (Chiroptera) often go unheeded and unnoticed because they are mainly active by night. Yet they represent the second largest order of mammals, totalling some 800 species, and in overall abundance they are unequalled.

Fossil remains indicate that bats had already evolved into a distinct order of mammals some 60 million years ago in the Eocene period. Their skeletal structure has changed very little in the intervening years and so it may be assumed that already in those far-off times they had achieved mastery over the air.

Bats range in size from the tiny pipistrelle (*Pipistrellus nanulus*), a native of West Africa, with a mass of no more than 2.5 g and a wing span of approximately 15 cm, up to the huge fruit bats (Pteropodidae). In this family there are several species which have a wing span of about 170 cm and a body mass approaching 1 kg.

A considerable amount is known about the flight apparatus of bats and the flight styles they adopt (Wimsatt, 1970).

8.1.1 The Bat Wing

The wing of a bat is totally different from that of a bird. It consists of a fold of skin which forms a membrane and is carried between the fore and rear limbs. The membrane is only capable of withstanding tensile loads. The leading edge of the wing is supported by the second digit, the proximal parts of the third digit, and a small membrane (dactylopatagium minus) which is carried between them, whilst the trailing edge of the wing is unsupported. The wing extends from the leading edge rearwards, the wing root consisting of the body and hind legs of the bat, whilst the elongated fourth and fifth digits play an important role in supporting aerodynamic loads and determining the contour of the wing at intermediate positions along the wing span. This structural arrangement is more efficient than that which evolved in the pterosaurs, which were unable to control the profile of the wing section, and has conferred upon the bats distinct aerodynamic advantages.

In particular, bats have been observed to increase the camber of the wing in the mid-span region by depression of the fifth digit, thereby raising the maximum lift coefficient generated by the wing. This effect is exploited under conditions of low-speed flight, and has an exact parallel in the world of aviation, where the flaps of an aircraft are extended from the trailing edge of the wing in order to assist flight at the low airspeeds appropriate to take-off and landing.

8.1.2 Aerodynamics

The primary parameters which govern the performance of the larger flying animals are the wing loading, W/S, and the aspect ratio, b/c. Bats encompass a

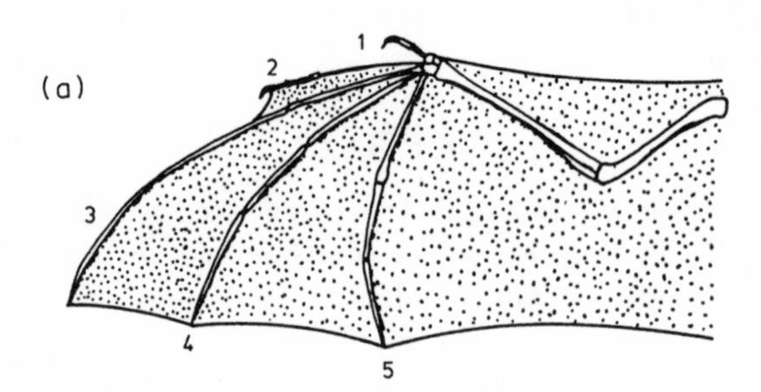

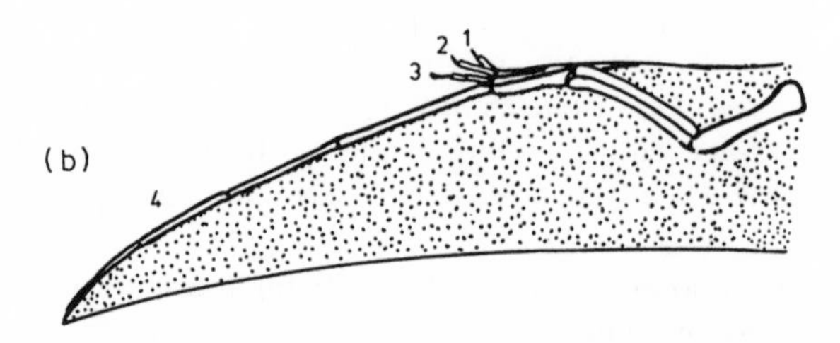

Figure 8.1. Comparison of the skeletal arrangement and flying surfaces of (a) bats and (b) pterosaurs. These may be compared with the human arm and bird wing, which are shown in Fig. 5.3.

range of wing loadings from about 7 to 70 N m^{-2}, the lower figure being appropriate to the very small bats, such as the pipestrelle, whilst the upper figure relates to the larger fruit bats. The wing geometries of bats cover aspect ratios from roughly 5 to 10. Morphological data on a number of species of bat are provided in Table 8.1.

Table 8.1 Morphological data on some bats.

Bat	Mass (g)	Wing span (cm)	Wing loading (N m^{-2})	Aspect ratio
Plecotus auritus (common long-eared bat)	9	27	7.2	5.9
Nyctalus noctula (noctule or great bat)	27	32	22.1	8.5
Otomops martiensseni (an African molossid bat)	35	47	15.6	10
Pteropus medius (an Old World fruit bat)	923	51.5	40	7.6

A comparison of these parameters with the corresponding figures for birds shows that there is considerable overlap. The range of aspect ratios appropriate to bats falls entirely within the range covered by birds (from approximately 4 to 18). So far as wing loading is concerned there is again considerable overlap, but with a vital difference. The wing loadings of the smaller bats descend to levels below any at which birds operate.

From these initial comparisons some immediate and crucial deductions can be made. Where there is overlap of the aerodynamic parameters a close correspondence between performances of birds and bats can be expected, both in respect of flight styles and as regards the detailed functioning of the wings as aerodynamic surfaces. These expectations of convergence are indeed borne out by observation. Hence, where such correspondence exists, in order to avoid undue repetition a rather brief treatment of the aerodynamics of the bat wing will be presented here and the reader is referred to the appropriate discussion of the aerodynamics of the bird wing for further details.

The smaller bats have been shown to operate at very low wing loadings. It is this factor, more than any other, which accounts for the one aspect of flight performance where bats are unequalled by birds, namely their manoeuvrability, particularly at low speeds.

The wing of a bat is essentially a thin aerofoil of high camber, significantly higher than that of a typical bird, and this property greatly facilitates flight under low-speed, high-lift conditions.

In steady horizontal flight the components of the wing inboard of the fifth digit—the propatagium and the plagiopatagium—are mainly responsible for the generation of aerodynamic lift, whereas that part of the wing outboard of the fifth digit—the chiropatagium—produces most of the thrust.

The wing is heavily loaded during the downstroke, but is only lightly loaded

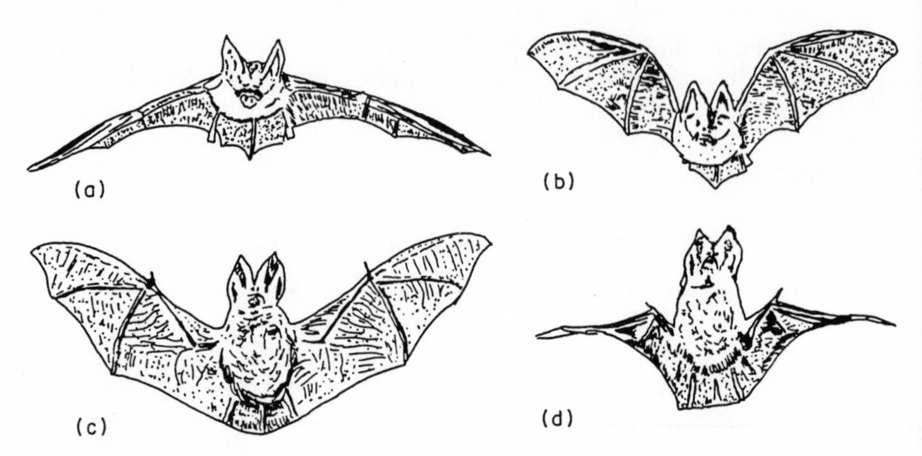

Figure 8.2. The long-eared bat (*Plecotus auritus*) in horizontal flight. Views (a) and (b) are from directly ahead; views (c) and (d) are the corresponding views from vertically below. The downstroke is shown in (a) and (c); the upstroke in (b) and (d).

during the upstroke and in this respect there are obvious similarities here between the mechanics of flight in birds and bats (Fig. 8.2).

For flight at low speeds the amplitude, but not the frequency, of the wingbeat is substantially increased. At the commencement of the downstroke the wings are raised with the wing tips positioned behind the centre of mass of the bat; the wings are fully extended and the membrane is stretched taut. During the power stroke the movement of the wing is both down and forward, whereas the recovery during the upstroke is attained through simultaneous upward and backward movement of the wing. There are great similarities between the wing movement adopted by the bat in this phase of flight, and the corresponding movement of the pigeon's wing depicted in Fig. 6.3.

An important aspect of the aerodynamics of chiropteran flight is the ability of bats to generate high lift at low airspeeds. The skeleton of bones which form the load-bearing structure of the wing produces appreciable irregularities to the contour of the wing close to the leading edge. Anyone unacquainted with aerodynamic phenomena might think that these features would have a deleterious effect on the performance of the wing as a lifting surface, but the opposite is in fact true. When a wing produces large amounts of lift, the pressure distribution generated on the upper surface is unfavourable to the continued development of an attached boundary layer, particularly towards the trailing edge. Separation of the boundary layer, if it were to occur, would result in the wing stalling, with a loss of lift and serious consequences; so this situation must be avoided at all costs. Turbulent boundary layers are much less prone to separation than are laminar layers, because of their ability to transfer momentum laterally throughout the boundary layer, due to the mechanism of turbulent mixing. In this sense it is aerodynamically advantageous for the wing surface to be covered by a turbulent boundary layer. The humerus, radius, and the second and third digits of the bat wing are seen to serve the important function of promoting an early transition within the boundary layer, so that its subsequent growth over almost the entire wing surface is as a turbulent layer, with the consequent benefits to which reference has been made.

8.1.3 Flight styles

Some bats, particularly the larger species of the Old World fruit bats (Pteropodidae), are capable of gliding over short distances, but none has acquired the ability which certain birds possess of extracting energy from the wind by soaring. Other Chiroptera, including a number of the free-tailed bats (Molossidae), are known to include spectacular dives in their repertoire of flight styles. But, overall, bats rely much more than do birds on continuous beating of the wings to remain airborne. Already reference has been made to the remarkable skills which numerous bats display in performing aerial manoeuvres. Amongst birds only certain of the flycatchers (Muscicapidae) remotely approach these feats of performance. Of the remaining flight styles employed by bats, parallels can be identified from the bird world, as a result of the processes of avian and chiropteran evolutionary convergence.

Slow, manoeuvring flight

The characteristic chiropteran style of flight, which most readily distinguishes bats from birds, is forward flight, at slow to moderate airspeeds, which is constantly interrupted by intricate aerial manoeuvres, involving sideslip and roll, and changes in flight direction. This type of flight is practised in its most highly developed form by the insectivorous bats, but some of the smaller nectar-feeding bats are also capable of flying in this manner, delicately avoiding objects, such as the branches of shrubs or trees, obstructing their flight path. Small insects often have extremely erratic flight patterns and the pursuit and capture of its food supplies is the motivation for the rapid aerial manoeuvres of the insectivorous bats. The low wing loading is a crucial advantage to this style of flight, because to be able to change directions rapidly, a bat must develop substantial transverse aerodynamic forces to overcome the inertial force tending to maintain the motion of the bat along a straight path.

Another important factor is the pronounced camber of the bat wing which facilitates the generation of high lift at low airspeeds. But these factors alone are insufficient: in order to manoeuvre rapidly, and particularly to deviate from a straight flight path, a bat must be able to modify the basic symmetry of its flight apparatus during the wingbeat cycle. This is where the flexibility of the bat's wing membrane and the ability to control its profile using the fifth digit is crucial, particularly as the musculature of the two wings can be operated separately. With one or two asymmetrical beats of the wing, a bat is able to execute a rapid sideslip, rolling manoeuvre, or change in flight direction.*

* Both wings continue to beat in phase.

The uropatagium is the membrane borne between the hind limbs of the bat. This aerodynamic surface is not essential to the performance of aerial manoeuvres, as its absence from species which are capable of executing rapid turns demonstrates. Nevertheless, there are species whose ability to manoeuvre is further enhanced by suitable use of this membrane. Insectivorous bats are highly skilled at catching their prey in flight. In some species insects are captured by 'cupping' the wing membrane and uropatagium beneath the body to form a trap.

Hovering

A number of species of bat are capable of hovering in still air without difficulty. They include amongst their number some of the insectivorous Chiroptera, but the ability to hover is most advanced amongst certain of the nectar-feeding bats, including members of the sub-families Macroglossinae and Glossophaginae. These Chiroptera take to the wing as darkness falls and they search for and gather nectar from the night-blooming plants of the tropics, with a nocturnal behaviour pattern which closely parallels the diurnal activity of the hummingbirds.

Whereas insectivorous bats spend long foraging periods on the wing, food acquisition by the nectar-feeding bats is accomplished by shorter, intermittent intervals of flight, as hovering is an intensely demanding activity in terms of energy expenditure.

The hovering, nectar-feeding bats tend to possess broad wings of low aspect ratio (about 5), a morphological characteristic which seems to be advantageous for hovering flight, but incurs some penalties as regards the aerodynamics of horizontal, sustained flight, due to the increased vortex drag.

Studies have been made of the hovering flight of certain bats. The long-eared bat (*Plecotus auritus*) is an example, and, under these conditions of flight, this bat beats its wings at a frequency of about 11 cycles per second. Analysis of this hovering flight suggests that lift coefficients between 3 and 6.5 would be necessary to sustain the bat in the air, if the bat relied solely upon techniques which were adaptations of steady aerodynamics. Such values are considerably in excess of those which occur under conditions of steady aircraft flight, and are indicative of the fact that bats, along with various birds and insects, utilize complex mechanisms of lift generation which can only be explained properly in terms of unsteady aerodynamics. The understanding of such matters is still at a rather rudimentary stage, and, since the principles are broadly the same for bats as for insects, the reader is referred to the discussion of unsteady aerodynamics contained in the chapter on insect flight.

Sustained horizontal flight

The larger bats, including most of the Old World fruit bats (Pteropodidae), have relatively high wing loadings and, because of the relationship between flight speed and wing loading (see section 5.3), these chiroptera exhibit a distinctive flight style, characterized by strong, direct horizontal flight, at a typical wingbeat frequency of 2.5–3 cycles per second. Appreciable distances, up to 20 km, may separate the roosts of these frugivorous bats from the areas where they feed, and so the purposeful style of flight they adopt is beneficial in making these regular cross-country journeys.

High-speed flight

Renowned for their fast flight are the tropical free-tailed bats (Molossidae). These bats have high aspect ratio (9 or 10) wings, a high wingbeat frequency, and their morphology, flight style, and rapid movement through the air make them the chiropteran equivalent of the swift.

The molossids possess a number of anatomical features which indicate a specialized adaptation for high-speed flight. At high speeds, profile drag, increasing as V^2, is far more important than vortex drag, and the body of the molossids is therefore streamlined and compact to minimize this component of drag. So far as the wing is concerned, the bones at the leading edge are of a stream-lined section and provide the wing with a smooth contour, in contrast to the wing section of bats designed for low-speed, high-lift conditions. The ears are small and aerodynamically efficient (in stark contrast to the huge ears which some of the bats which operate at low speeds possess; at low speeds vortex drag is the major component of drag). Thrust production is vital at high speeds, and the wings of the molossids are well adapted for this purpose. They are of high aspect ratio and are less highly cambered than the wings of other bats. The wing membranes are tougher, more leathery, and less flexible than in other species of bat. Special adaptations are to be found in the musculature of the wings of the mollosids and in the presence of a network of elastic fibres, which, upon contraction of the muscles, become taut and give the wing increased rigidity over the entire span.

The high wingbeat frequency, high aspect ratio, and the ability to operate the wing with considerable rigidity are indicative of a convergence of the molossid bats, from the chiropteran world, and the swifts, from the avian world, upon a common solution to the problem of sustained high-speed flight.

Like the swift, some of the molossids find landing and take-off difficult, and they employ the same solution to the problem, trading kinetic energy for potential energy.

8.2 THE PTEROSAURS

There was a period in history when several types of flying reptile, now extinct, took to the air. These animals, the pterosaurs, evolved independently of, and are unrelated to, the bats and birds, creatures with which they are most readily compared because of their life styles. A number of skeletons of these early animals have survived, and from these remains it has proved possible to deduce a great deal about their mode of existence.

The earliest of the pterosaurs appeared in the Triassic period, 200 million years ago. An example of this early type is the Rhamphorhynchus, which was a small animal, about 60 cm from nose to tail. Like all pterosaurs, it had a narrow pointed wing formed by a fold of skin, carried between the fore- and hind-limbs. The first three digits of the fore-limb were used as claws, whilst the fourth was enormously elongated and formed the leading edge of the wing over the majority of the span. This pterosaur had a long stiff tail and the feet were webbed.

Projecting forward there was a long slender neck which carried a large head with a pointed jaw containing numerous teeth. All the larger bones of the skeleton were hollow, as they are in birds. With a wing span of about 120 cm and a mass of about 200 g, Rhamphorhynchus had a wing loading of approximately 60 N m^{-2}.

All the early pterosaurs resembled Rhamphorhynchus, possessing long tails with a horizontal surface. This assisted the longitudinal stability of the animal, but had the disadvantage of restricting the range of possible flight speeds, by limiting the incidence angle over which the wing could be used. As a result of evolutionary pressures, in the course of time tailless forms of pterosaurs appeared and the forms with tails gradually became extinct. A whole range of tailless pterosaurs existed, the smallest being similar in size to a sparrow, whilst at the opposite extreme the Pteranodon had a wing span extending to nearly 7 m (23 ft), over double that of the bird with the largest span, the wandering albatross.

A considerable amount has been learnt about the Pteranodon. Its mass was approximately 16.5 kg (36.5 lb), its wing area was 4.6 m^2 (50 ft^2), the aspect ratio was 10.5 and it had a low wing loading of about 35 N m^{-2}. Thus the Pteranodon was heavier than any species of bird which flies today. The form of aerial locomotion usually adopted by Pteranodon was that of gliding. Tests on models (Bramwell and Whitfield, 1977) suggest that its optimum gliding performance occurred at airspeeds of about 8 m s^{-1} (18 mph), under which conditions its wings operated at a lift/drag ratio of 18.5 and the sinking speed was as low as 0.42 m s^{-1} (0.94 mph). Careful examination of the skeletal remains of Pteranodon has also revealed that the wings were capable of being flapped. The weight of Pteranodon was 16.6×9.81 N and so, at the minimum sinking speed of 0.42 m s^{-1}, the rate of energy expenditure was 68 W. In order to sustain itself in level flight the wing muscles would have had to supply power in excess of this figure. Consideration of the bone structure of Pteranodon shows that it would have been able to carry wing muscles of mass up to about 3 kg, representing about 20 per cent of body weight, which is consistent with the musculature of many birds. To avoid overloading the bones in the wing, the Pteranodon would have been restricted to a very low wingbeat frequency, and a figure of one complete cycle every 2 seconds might be representative. Thus the flight muscles would have been required to contract at this same frequency. Measurements with birds indicate an approximate upper limit to the energy release per muscle contraction as 65 J kg^{-1}, from which it is deduced that the maximum power output of Pteranodon's muscles would have been about 97.5 W.

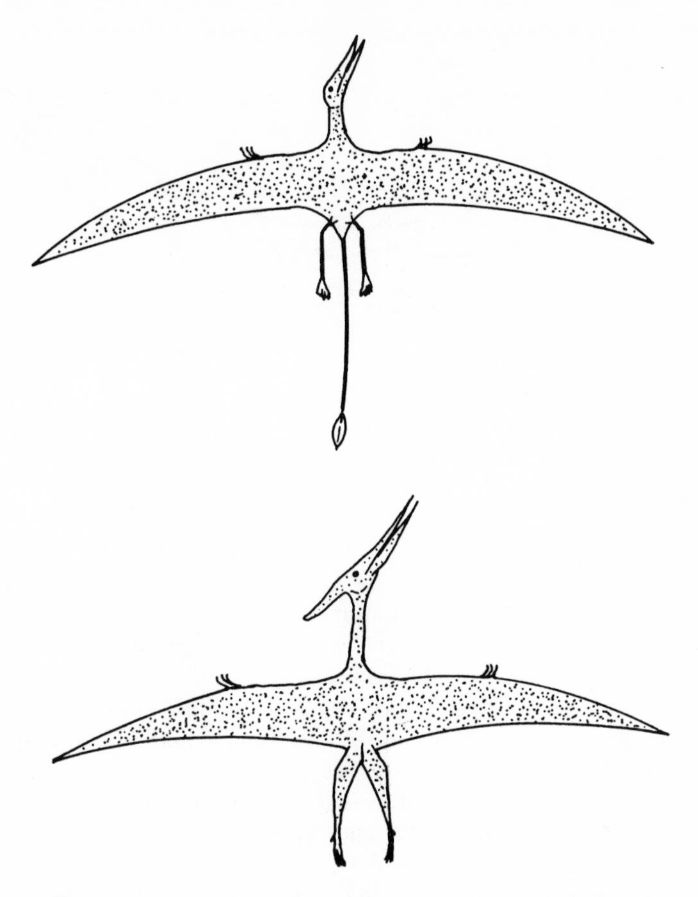

Figure 8.3. The two sub-orders of Pterosauria. The Rhamprorhynchoidea (top) had a long tail and a long neck. The Pterodactyloidea (bottom) had no tail, whilst the neck and face were very elongated. Both sub-orders had high aspect-ratio wings.

Bearing in mind the crude estimates that have had to be made for muscular mass and wingbeat frequency, these calculations reveal that Pteranodon was probably just about capable of sustained straight and level flight, but there was little margin available for any manoeuvres involving climbing or turning. Take-off and landing would have presented particular difficulties, because the wings would have been quite incapable of withstanding the sorts of loads that many birds accept under conditions of slow forward flight. In several respects the limitations imposed on Pteranodon by its morphology have their parallel today in the wandering albatross. Just how closely their life styles match is, however, a matter of conjecture. We have no means of knowing, for example, whether the pterosaurs could exploit the flight style of dynamic soaring, of which the albatross is a master. The chances are probably against this having happened.

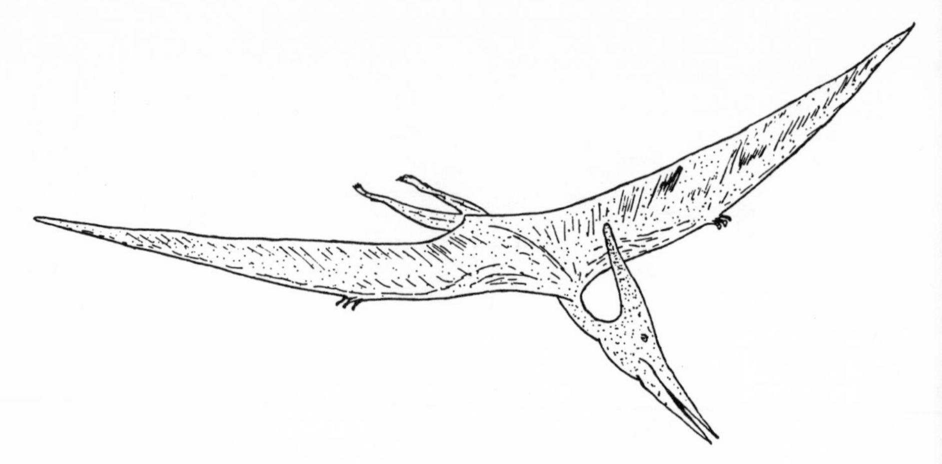

Figure 8.4. Impression of Pteranodon in gliding flight.

Until 1971 it was generally considered that Pteranodon, with a wing span of some 7 metres, represented the maximum size of any flying animal that had inhabited the earth. However, in that year a number of wing bones from a previously unknown species of pterosaur, subsequently named *Quetzalcoatlus northropi*, were discovered in West Texas, USA. The radius, a forearm bone, was essentially complete, however, and larger than the comparable bone in other pterosaurs. It was not possible to reconstruct the complete wing from these fossil remains. By scaling the dimensions of pterosaurs with similar proportions it has been deduced that *Quetzalcoatlus* had a wing span of between 11 and 12 metres. Using the scaling law $m \propto l^3$, it is estimated that such an animal would have had a mass of about 86 kilograms. Since a substantial amount of extrapolation, with a considerable degree of uncertainty, was required to obtain the estimates of both wing span and body mass, until further supportive evidence becomes available these data must be treated with considerable caution. It is possible, but perhaps improbable, that *Q. northropi* was the largest creature to fly; as yet the case is unproven.

Clearly the continued survival of Pteranodon and the larger pterosaurs was always on a knife-edge, but the smaller pterodactyls could resort to powered flight with a greater margin between the required and available powers. For reasons which are not yet clear, however, the pressures for survival placed upon pterosaurs eventually became too severe and they finally became extinct by the end of the Cretaceous period, 70 million years ago. A possible explanation for the decline of the pterosaurs was climatic change. To survive and prosper, with their limited aerodynamic capability, pterosaurs required calm conditions or, at worst, light breezes. It is thought that in those far-off days, there came a time when climatic changes were accompanied by greater seasonal variations in the weather and in particular the occurrence of stormy periods. Unable to adapt to these conditions the pterosaurs perished.

8.3 GLIDING ANIMALS

A number of animals, whilst not having risen to the supreme form of active flying, have nevertheless developed the ability to glide (Fig. 8.5).

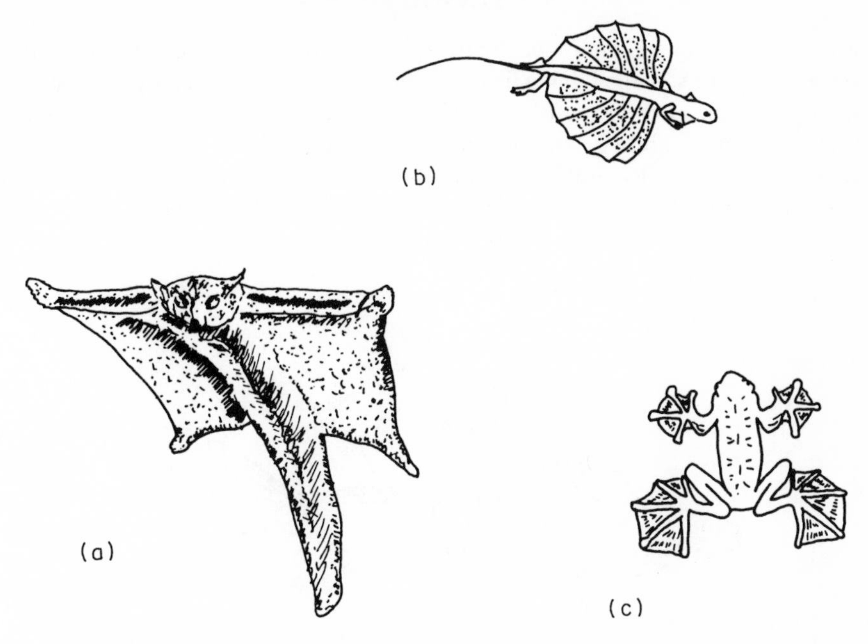

Figure 8.5. Some animals capable of gliding: (a) the cobego or flying lemur (*Cynocephalus volans*); (b) the gliding lizard (*Draco volans*); (c) the flying frog (*Rhacophorus rheinwardtii*)

This particular flight style is adopted by a range of arboreal mammals which have evolved independently. All of these creatures have a fold of skin, known as the patagium, which stretches between the fore- and hind-limbs, and when these are spread out laterally the patagium acts as an efficient aerodynamic surface, generating substantial lift. Mammals exploiting this means of aerial locomotion are two species of cobegos (*Cynocephalus*), often referred to as flying lemurs, the flying squirrels of the family Sciuridae and the scaly-tails of the family Anomaluridae, which are both rodents, and the flying phalangers of the family Phalangeridae, which are marsupials.

The gliding lizards *Draco volans* and *Draco maculatus* have also evolved an efficient aerodynamic surface. In contrast to the mammals the legs are not involved in creating the flying surface; instead this is supported by a series of ribs which are extended laterally from the side of the body. These creatures are capable of gliding over distances of 40–50 m, with a loss in height of about 1 in 4.

The predominant aerodynamic force generated by these gliding animals is that of lift. Inevitably they also experience drag in the forms of profile drag and vortex drag. The angle of the glide path relative to the horizontal is, for small glide angles, equal to the lift/drag ratio, so the facility to bridge the gap between neighbouring trees is enhanced with increasing lift/drag ratio.

8.4 FLYING FISHES

Final interesting examples of creatures which have evolved efficient lifting surfaces are the so-called flying fishes. In the family of marine fishes Exocoetidae the pectoral fins are substantially enlarged and these surfaces are exploited for the purposes of gliding. In certain species, for example those of the genus *Exocoetus*, only the pectorals are enlarged. In others, for example *Cypsilurus*, the pelvic fins are enlarged as well (Fig. 8.6). The flight of these fishes is achieved in

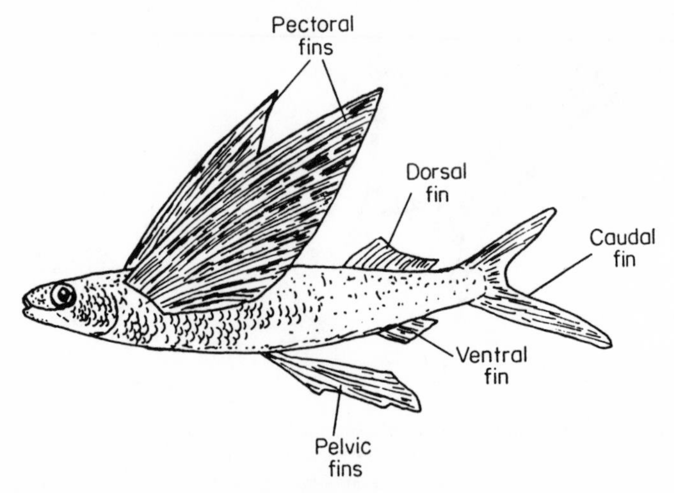

Figure 8.6. The flying fish *Cypsilurus heterusus*, which grows to a length of about 30 cm.

the following way. First of all the fish attains a high speed under water, with the flying fins folded, and the momentum so gained is used to break through the surface of the sea. As this happens, first the pectoral fins and then the pelvics are spread, only the lower lobe of the tail being maintained in contact with the sea. This lobe is vigorously moved from side to side, and the velocity of the fish increases still more as it follows a zig-zag course over the surface. Eventually when sufficient speed has been attained the fish breaks clear of the surface and becomes fully airborne. It holds the fins in a fixed configuration and the aerodynamic forces derived on these surfaces allow the fish to glide through the air over a considerable distance. In extreme cases they are capable of covering up to 100 m (well over 100 yd) and rise to heights of 6–9 m (20–30 ft). Occasionally they have been known to land on the deck of a passing ship.

Amongst freshwater fishes the South American hatchet fishes (*Gaster opelecidae*) are the most efficient flyers. As the name of these fishes indicates, they have a body with a deep keel and this is maintained in contact with the water as the fish skims over the surface. The pectoral fins are vibrated rapidly, producing a distinctive whirring noise, and they generate sufficient aerodynamic lift for the fishes to make short excursions into the air. The freshwater butterfly fish *Pantodon*, of Africa, is similarly capable of beating its pectoral fins and also makes flights of short duration.

References

CHAPTER 1

Allen, J. E. (1982). *Aerodynamics: The Science of Air in Motion*. Granada, London.
Houghton, E. L., and Carruthers, N. B. (1982). *Aerodynamics for Engineering Students*, 3rd edition. Edward Arnold, London.
Schlichting, H. (1979). *Boundary-layer Theory*, 7th edition. McGraw-Hill, New York.
Shapiro, A. H. (1961). *Shape and Flow*. Heinemann, London.
Sutton, O. G. (1949). *The Science of Flight*. Penguin Books, London.

CHAPTER 2

Clift, R., Grace, J. R., and Weber, M. E. (1978). *Bubbles, Drops and Particles*, Academic Press, New York.
Mason, B. J. (1978). Physics of a raindrop. *Phys. Educ.*, **13**, 414–419.
Schlichting, H. (1979). *Boundary-layer Theory*, 7th edition. McGraw-Hill, New York.

CHAPTER 3

Achenbach, E. (1974). The effects of surface roughness and tunnel blockage on the flow past spheres. *J. Fluid Mech.*, **65**, Pt 1, 113–125.
Bailey, I. H., and Macklin, W. C. (1968). The surface configuration and internal structure of artificial hailstones. *Q. J. R. Meteorol. Soc.*, **94**, 1–11.
Clift, R., Grace, J. R., and Weber, M. E. (1978). *Bubbles, Drops and Particles*. Academic Press, New York.
Goldie, E. C. W., Meaden, G. T., and White, R. (1976). The concentric halo display of 14 April 1974. *Weather*, **31**, 61–69
Greenler, R. (1980). *Rainbows, Halos and Glories*. Cambridge University Press.
Greenler, R. G., Mueller, J. R., Hahn, W., and Mallmann, A. J. (1979). The 46° Halo and its arcs. *Science*, **206**, 643–649.
Hobbs, P. V. (1974). *Ice Physics*. Oxford University Press.
Ludlam, F. H. (1980). *Clouds and Storms*. Pennsylvania State University Press.
Macklin, W. C., and Ludlam, F. H. (1961). The fall-speeds of hailstones. *Q. J. R. Meteorol. Soc.*, **87**, 72–81.
Roos, F. W., and Willmarth, W. W. (1971). Some experimental results on sphere and disk drag. *American Institute for Aeronautics and Astronautics Journal*, **9** (2), 285–291.
Schlichting, H. (1979). *Boundary-layer Theory*, 7th edition. McGraw-Hill, New York.
Willmarth, W. W. (1964). Steady and unsteady motions and wakes of freely falling disks. *Physics of Fluids*, **7** (2), 197–208.

CHAPTER 4

Burrows, F. M. (1975). Wind-borne seed and fruit movement. *New Phytol.*, **75**, 405–418.
Houghton, E. L., and Brock, A. E. (1960). *Aerodynamics for Engineering Students.* Edward Arnold, London.
McCutchen, C. W. (1977). The spinning rotation of ash and tulip tree samaras. *Science*, **197**, 691–692.
Norberg, R. A. (1973). Autorotation, self-stability, and structure of single-winged fruits and seeds (samaras) with comparative remarks on animal flight. *Biol. Rev.*, **48**, 561–596.

CHAPTER 5

Hankin, E. H. (1913). *Animal Flight.* Iliffe and Sons, London.
Lighthill, M. J. (1974). Aerodynamic aspects of animal flight. *Bull. Inst. Maths Appl.*, **10**, 369–393.
Pennycuick, C. J. (1972). *Animal Flight.* Edward Arnold, London.
Spillman, J. (1978). The use of wing tip sails to reduce drag. *Aeronaut. J.*, **82**, 387–395.
Tucker, V. A. and Parrott, G. C. (1970). Aerodynamics of gliding flight in a falcon and other birds. *J. Exp. Biol.*, **52**, 345–367.

CHAPTER 6

Eastwood, E. (1967). *Radar Ornithology.* Methuen, London.
Greenewalt, C. H. (1962). Dimensional relationships for flying animals. *Smithsonian Misc. Coll.*, **144** (2).
Greenewalt, C. H. (1975). The flight of birds. *Trans. Am. Phil. Soc.*, **65** (4), 1–67.
Hingston, R. W. G. (1927). Bird notes from the Mount Everest expedition of 1924. *J. Bombay Nat. Hist. Soc.*, **32**, 320–329.
Hummel, D (1983). Aerodynamic aspects of formation flight in birds. *J. Theor. Biol.*, **104**, 321–347.
Lighthill, M. J. (1974). Aerodynamic aspects of animal flight. *Bull. Inst. Maths Appls.*, **10**, 369–393.
Lighthill, M. J. (1977). Introduction to the scaling of aerial locomotion. In T. J. Pedley (ed.), *Scale Effects in Animal Locomotion*, pp. 365–404. Academic Press, New York.
Lissaman, P. B. S., and Shollenberger, C. A. (1970). Formation flight of birds. *Science*, **168**, 1003–1005.
May, R. M. (1979). Flight formations in geese and other birds. *Nature*, **282**, 778–780.
Meinertzhagen, R. (1955). The speed and altitude of bird flight. *Ibis*, **97**, 81–117.
Norberg, U. M. (1974). Hovering flight in the pied flycatcher (*Ficedula hypoleuca*). In T. Y. Wu, C. J. Brokaw, and C. Brennen (eds), *Swimming and Flying in Nature*, vol. 2, pp. 869–881. Plenum Press, New York.
Pennycuick, C. J. (1972). *Animal Flight.* Edward Arnold, London.
Rayner, J. M. V. (1977). The intermittent flight of birds. In T. J. Pedley (ed.), *Scale Effects in Animal Locomotion,* pp. 437–443. Academic Press, New York.
Rayner, J. M. V. (1979). A vortex theory of animal flight. Part 1. The vortex wake of a hovering animal. *J. Fluid Mech*, **91** (4), 697–730. Part 2. The forward flight of birds, *J. Fluid Mech.*, **91** (4), 731–763.
Tyne, J. V., and Berger, A. J. (1959). *Fundamentals of Ornithology.* John Wiley, New York.
Vaurie, C. (1972). *Tibet and its Birds.* Witherby, London.
Ward-Smith, A. J. (1984). Analysis of the aerodynamic performance of birds during bounding flight. *Math. Biosciences*, **68**, 137–147.

CHAPTER 7

Harris, J. R. (1956). *An Angler's Entomology*. Collins, London.
Imms, A. D. (1971). *Insect Natural History*, 3rd edition. Collins, London.
Kuethe, A. M. (1975). On the mechanics of flight of small insects. In T. Y. Wu, C. J. Brokaw, and C. Brennen (eds), *Swimming and Flying in Nature*, vol. 2, pp. 803–813. Plenum Press, New York.
Langmuir, I. (1938). The speed of the deer fly. *Science*, **87**, 233–234.
Lighthill, M. J. (1973). On the Weis-Fogh mechanism of lift generation. *J. Fluid Mech.*, **60** (1), 1–17.
Mani, M. S. (1962). *Introduction to High Altitude Entomology*. Methuen, London.
Norberg, R. A. (1975). Hovering flight of the dragonfly *Aeschna juncea*—kinematics and aerodynamics. In T. Y. Wu, C. J. Brokaw, and C. Brennen (eds), *Swimming and Flying in Nature*, vol. 2, pp. 763–781. Plenum Press, New York.
Savage, S. B., Newman, B. G., and Wong, D. T-M. (1979). The role of vortices and unsteady effects during the hovering flight of dragonflies. *J. Exp. Biol.*, **83** (1), 59–77.
Scorer, R. S. (1978). *Environmental Aerodynamics*. Ellis Horwood, Chichester, England.
Townsend, C. H. T. (1927). Cephenomyia. *J. New York Ent. Soc.*, **35** (3), 250–251.
Weis-Fogh, T. (1973). Quick estimates of flight fitness in hovering animals, including novel mechanisms for lift production. *J. Exp. Biol.*, **59**, 169–230.

CHAPTER 8

Bramwell, C. D., and Whitfield, G. R. (1974). Biomechanics of Pteranodon. *Phil. Trans. R. Soc.*, Series B, **267**, 503–581.
Wimsatt, W. A. (1970). *Biology of Bats*, vol. 1. Academic Press, London.

List of Symbols

This list is not exhaustive; it is intended to cover the main symbols used throughout the text. Where a symbol has more than one meaning the context provides clarification.

A	aspect ratio, b^2/c
A_D	disc area
b	wing span
C_D	drag coefficient
C_{D0}	zero-lift drag coefficient
C_L	lift coefficient
c	wing mean chord
D	drag force
d	diameter
e	energy content of body fat per unit mass
g	acceleration due to gravity
h	dimension of ice crystal along the principal axis
K	lift-dependent drag factor
k	height of surface roughness element
L	aerodynamic lift
l	characteristic length, or length of columnar crystal
m	mass
P	power
p	pressure
R	gas constant, for air $R = 287\,\mathrm{J\,kg^{-1}\,K^{-1}}$, or range
Re	Reynolds number
r	radial distance
S	reference area
s	semi-span of wing
T	thrust or temperature, Kelvin
t	thickness of plate crystal, or temperature, °C
u	velocity component in streamwise direction
$\mathcal{V}$	volume
V	velocity of body through the air, or velocity of airstream past a stationary body

V_a airspeed
V_{bg} best glide speed
V_c cruising speed
V_g ground speed
V_{md} minimum drag speed
V_{mp} minimum power speed
V_{ms} minimum sink speed
V_s stalling speed
V_T terminal velocity
V_W wind speed
v velocity component in vertical direction
v_s sinking speed, measured in the vertical direction
W weight
w width across the corners of ice crystal, or velocity component in transverse direction
y coordinate normal to the streamwise direction
z altitude

α angle of incidence
β coning angle
ε angle of glide path relative to the horizontal
η efficiency
λ atmospheric temperature gradient,
μ dynamic viscosity
ν kinematic viscosity
ρ air density
σ mean density of particle
τ shear stress
ϕ angle of bank measured relative to vertical
ω angular velocity, or wingbeat frequency

Subscript
n nth layer of standard atmosphere

Glossary of Terms

An effort has been made to use definitions which are the same as, or are consistent with, those in British Standard BS185, but occasional departures from that scheme have been made in the interests of simplicity.

Actuator disc	A concept in the momentum theory of rotors, in which the rotor is treated as equivalent to an infinite number of elementary aerofoils capable of producing a discontinuous, uniformly distributed pressure rise.
Aerodynamic centre	The point about which the rate of change of pitching moment with incidence is zero.
Aerodynamics	The branch of physics which deals with the movement of a body through the air, or the motion of air about a stationary body.
Aeroelasticity	The branch of physics which deals with the phenomena resulting from aerodynamic, inertial, and elastic forces.
Aerofoil	A body so shaped as to produce an aerodynamic force perpendicular to its direction of motion through the air, without incurring excessive drag.
Aerofoil section	The shape of the boundary of a section of an aerofoil in a plane parallel to its plane of symmetry.
Aeronautics	All activities relating to aerial locomotion.
Aeroplane	A power-driven heavier-than-air aircraft with lift-generating surfaces which remain fixed under given conditions of flight.
Aircraft	A vehicle designed to travel through the air outside the ground effect region.
Airspeed	The speed of an object relative to the ambient undisturbed air.
Altitude	The vertical distance above mean sea level.
Angle, gliding	The angle between the flight path in still air and the horizontal.

Angle of incidence	The angle between a reference line in a body (for an aerofoil the chord is usually used) and the direction of motion of the airstream.
Aspect ratio	The ratio of the square of the span to the gross area of an aerofoil; or the ratio of the span to the mean chord.
Atmosphere	The air surrounding the earth.
Atmospheric pressure	The pressure produced at any point in the atmosphere by the weight of the air above it.
Autorotation	Continuous rotation of a body resulting from the action of aerodynamic forces.
Ballistic trajectory	The path followed by a body under gravitational and aerodynamic forces only.
Boundary layer	The thin layer of fluid adjacent to a surface, in which the viscous forces are dominant.
Bluff body	Any body whose shape lacks streamlining.
Camber	(1) Curvature of the camber line of an aerofoil section. (2) The ratio of the maximum height of the camber line above the chord to the chord length.
Camber line	A line each point of which is equidistant from the upper and lower boundaries of the aerofoil section, the distances being measured normal to the chord.
Centre of pressure	The point on some reference line (e.g. the chord of an aerofoil) about which the pitching moment is zero.
Circulation	The integral of the component of the fluid velocity along any closed path with respect to the distance round the path.
Chord	The straight line through the centres of curvature at the leading and trailing edges of an aerofoil section.
Chord, mean	A chord of length equal to the gross wing area divided by the span.
Dihedral	The angle at which the port and starboard parts of the wing are inclined upwards relative to the transverse plane of reference.
Disc loading	The aerodynamic force generated by the rotor divided by the area of the circle described by the rotating motion of the tips of the rotor.
Dive	A steep descent.
Downwash	The vertical (downward) component of velocity which is found in the flow field about an aerofoil and is a consequence of the lift generated by the aerofoil.

Drag — The component of the total aerodynamic force which acts on a body in the same direction as the velocity vector defining the relative motion of that body and the air, and acts to resist that motion.

Drag coefficient — A dimensionless quantity which is a measure of the air resistance associated with the movement of a body through the air; see equation (1.13).

Drag, form — The normal-pressure drag less the vortex drag.

Drag, lift-dependent — The difference between the drag at a given lift coefficient and the drag at some specified datum lift coefficient.

Drag, normal pressure — Drag arising from the resolved components of the pressures normal to the surface of a body.

Drag, profile — The sum of the form drag and the skin-friction drag.

Drag, skin-friction — Drag arising from the resolved components of the tangential forces on the surface of a body.

Drag, vortex — Drag arising from the formation of trailing vortices (formerly called induced drag).

Fluid — A general term which embraces all liquids and gases, including air.

Glider — A non-power-driven heavier-than-air aircraft.

Ground effect — The effect of the proximity of the ground or other surface on the aerodynamic characteristics of a moving body. Hovercraft exploit this effect.

Groundspeed — The horizontal component of a moving body's velocity relative to the earth's surface.

Gust — A rapid variation with time or distance in the speed or direction of the wind.

Hovering — Remaining stationary relative to the ground in still air.

Hovering, wind-assisted — Remaining stationary relative to the ground by flying forward into the wind at the same speed as the wind moves relative to the ground.

Inviscid — Without viscosity.

Laminar flow — Flow in which there is no mixing of adjacent layers of fluid (except on a molecular scale).

Leading edge — The forward edge of an aerofoil or other body moving through the air.

Lift — The component of the total aerodynamic force in the direction perpendicular to the drag force.

Lift coefficient — A dimensionless quantity which is a measure of the lift generated by an aerofoil; see equation (1.12).

Moment — The product of a force and the length of the radius arm, at right angles to the line of application of the force, about a point. Unless opposed by an equal and opposite effect, a moment tends to give rise to rotational motion.

Parachute — An umbrella-shaped aerodynamic device to produce drag, commonly used to reduce the speed of a moving object.

Pitching moment — A moment tending to give rise to angular motion about a transverse axis.

Potential flow — Flow of an ideal, inviscid fluid in which the vorticity is zero everywhere.

Precipitation — A general term for the forms in which water may fall from the atmosphere.

Propeller — A power-driven assembly of aerofoils, radially disposed, designed to produce thrust by its rotation in air.

Quarter-chord line — The line through the quarter-chord points of an aerofoil.

Quarter-chord point — The point on the chord of an aerofoil section at one-quarter of the chord length behind the leading edge.

Range — The distance an aircraft (or bird) can travel under given conditions without refuelling.

Reynolds number — The product of a typical length and airspeed divided by the kinematic viscosity of the air. It expresses the ratio of the inertial forces to the viscous forces present within a fluid flow.

Rotor — A system of rotating aerofoils.

Sailplane — A glider designed to utilize only atmospheric currents for sustained free flight.

Samara — A winged fruit or seed, capable of generating aerodynamic lift.

Scale effect — The effect of a change in size.

Separation — Detachment of the flow from a solid surface with which it has been in contact.

Sinking speed — The vertical component of the airspeed during gliding flight.

Solidity — The ratio of the total blade area of a rotor to the area of the circle swept out by the rotor.

Span — The distance separating the wing tips.

Stability — The quality whereby any disturbance of steady motion tends to decrease. A given type of motion is stable if, following a disturbance, that motion is reassumed without movement of the flying surfaces.

Stall — (1) The progressive breakdown of attached flow over an aerofoil leading to the loss of lift.

	(2) To bring an aircraft into the flight condition in which the wings are stalled.
Standard atmosphere, international	A model atmosphere, adopted internationally for the purpose of comparing the performance of aircraft, the physical properties of which are specified as a function of altitude.
Streamline	An imaginary line drawn within a flow such that the local velocity vector is a tangent to the line. As a consequence of this definition no fluid flows across a streamline.
Streamlined shape	A body profile which has the property of low form drag.
Sweep-back	The angle in plan between a specified spanwise line along an aerofoil and the normal to the plane of symmetry. For an aerofoil as a whole, the quarter-chord line is preferred, but any other specified line, such as the leading edge, may be taken for a particular purpose.
Terminal velocity	The velocity for which the drag of a freely falling body just balances its weight at a given altitude.
Thickness/chord ratio	The ratio of the maximum thickness of an aerofoil section, measured perpendicular to the chord, to the chord length.
Thrust	The aerodynamic force attributed to the propulsive system.
Trailing edge	The rear edge of an aerofoil or other body moving through the air.
Transition	The change from laminar to turbulent flow.
Turbulent flow	Flow in which irregular fluctuations with time are superposed on a mean flow.
Twist, geometric	Variation, along the span of a wing or other aerofoil, of the angle between the chord and a fixed datum.
Unsteady	Changing with time.
Upwash	An upward component of velocity associated with the flow field of a trailing vortex.
Velocity, free stream	The velocity of the undisturbed fluid relative to a body immersed in it.
Viscosity	The quality which a fluid possesses of resisting shearing motion due to internal shear stresses.
Viscosity, dynamic	A physical property of a fluid which quantifies the effect of viscosity. It is equal to the product of the kinematic viscosity and the density of the fluid.
Viscosity, kinematic	A physical property of a fluid which quantifies the effect of viscosity. In the International System (SI)

	it has the units $m^2 s^{-1}$.
Vortex	A rotating body of fluid
Vortex, bound	A vortex used as a simplified representation of an aerofoil; its strength is chosen to give a lift equal to that generated by the aerofoil it replaces.
Vortex, starting	A vortex left behind when a lifting body is set in motion.
Vortex, horseshoe	A vortex consisting of a bound vortex transverse to the airstream with a trailing vortex springing from each of its ends. A system of horseshoe vortices is used for the purpose of analysis in aerofoil theory.
Vortex, trailing	A vortex arising at the tip of an aerofoil owing to the flow of air around the tip from the high-pressure region to the low-pressure region, and extending downstream from the wing tips.
Vortex street	A regular arrangement of vortices in two approximately parallel rows, which is sometimes formed behind cylindrical bodies.
Vorticity	Generally, rotational motion in a fluid, defined, at any point in the fluid, as twice the angular velocity of a small element of fluid surrounding the point.
Wake	The disturbed region of fluid behind a body, resulting from the passage of that body through the air.
Wind tunnel	An apparatus for producing a controlled stream of air for the purpose of conducting aerodynamic experiments.
Wing	The main supporting surface of an aircraft or flying animal.
Wing area, gross	The area of the surface bounded by the two wing tips and the leading and trailing edges continued to intersect in the plane of symmetry.
Wing area, net	The combined area of the two isolated surfaces bounded by the wing tip, the leading and trailing edges, and the wing root.
Wing loading	Weight divided by wing area.
Wing root	The line forming the intersection of the wing and body.
Wing tips	Those parts of the wing remote from the axis of symmetry.

Further Reading

CHAPTER 1

Allen, J. E. (1982). *Aerodynamics: The Science of Air in Motion.* Granada, London.
Houghton, E. L., and Carruthers, N. B. (1982). *Aerodynamics for Engineering Students*, 3rd edition. Edward Arnold, London.
Shapiro, A. G. (1961). *Shape and Flow.* Heinemann, London.
Simons, M. (1978). *Model Aircraft Aerodynamics.* Model and Allied Publications, Watford, England.
Sutton, O. G. (1949). *The Science of Flight.* Penguin Books, London.

CHAPTERS 2 AND 3

Clift, R., Grace, J. R., and Weber, M. E. (1978). *Bubbles, Drops and Particles*, Academic Press, New York.
Greenler, R. (1980). *Rainbows, Halos and Glories.* Cambridge University Press.
Hobbs, P. V. (1974). *Ice Physics.* Oxford University Press.
Ludlam, F. H. (1980). *Clouds and Storms.* Pennsylvania State University Press.
Minnaert, M. (1954). *The Nature of Light and Colour in the Open Air.* Dover Press, New York.
Pruppacher, H. R., and Klett, J. D. (1978). *Microphysics of Clouds and Precipitation.* D. Reidel Publishing Co., Dordrecht, Holland.
Scorer, R. S. (1978). *Environmental Aerodynamics.* Ellis Horwood, Chichester, England.
Tricker, R. A. R. (1970). *An Introduction to Meteorological Optics.* American Elsevier, New York.

CHAPTER 4

Gregory, P. H. (1961). *The Microbiology of the Atmosphere.* Leonard Hill (Books), London.
Ridley, H. N. (1930). *The Dispersal of Plants Throughout the World.* Reeve, Ashford, Kent.

CHAPTERS 5 AND 6

Pedley, T. J. (1977). *Scale Effects in Animal Locomotion.* Academic Press, London.
Pennycuick, C. J. (1972). *Animal Flight.* Edward Arnold, London.
Wu, T. Y-T, Brokaw, C. J., and Brennen, C. (1975). *Swimming and Flying in Nature.* Plenum Press, New York.

CHAPTER 7

Ellington, C. P. (1984). *The aerodynamics of hovering insect flight.* Phil. Trans. R. Soc. (**8**), 305, 1–181.

Nachtigall, W. (1974). *Insects in Flight.* Allen & Unwin, London.

Rainey, R. C. (1976). *Insect Flight.* Blackwell Scientific Publications, Oxford.

CHAPTER 8

Allen, G. M. (1982). *Bats.* Dover Publications, New York.

Wimsatt, W. A. (1970). *Biology of Bats*, vol. 1, Academic Press, London.

GENERAL

Alexander, R. McN. (1982). *Locomotion of Animals*, Chapter 4: Flight. Blackie, Glasgow.

Hertel, H. (1966). *Structure, Form, Movement.* Reinhold Publishing, New York.

Rayner, J. M. V. (1981). Flight adaptations in vertebrates. In M. H. Day (ed.), *Vertebrate Locomotion*, pp. 137–172. Academic Press, London.

Index

actuator disc theory, 63, 105, 121, 161
aerodynamic drag (*see* drag)
aerodynamic lift (*see* lift)
aeroelasticity, 81, 161
aerofoil, 161
  cambered, 15, 74, 145
  flat-plate, 14–15, 57
  section, 14, 161
aeroplane, 10, 72, 161
air
  density, 6, 99
  dynamic viscosity, 31
  gas constant, 3
  kinematic viscosity, 31
airspeed, 84, 89, 161
albatross, 70, 72, 88–91, 151
alula, 70, 98
angle
  glide, 59, 79
  incidence, 14, 15, 57, 58, 61, 73, 95, 162
archaeopteryx, 69
aspect ratio, 162
  bat wing, 145
  bird wing, 72, 73, 81, 84, 91, 145
atmosphere, standard, 3–5, 165
autorotation, 59–66, 162

bats, 10, 23, 143–149
Bernoulli's equation, 13, 17, 64
best glide speed, 79
bird
  fastest flying, 102
  heaviest flightless, 102
  heaviest flying, 102
  highest flying, 116
  longest migration, 112
bird flight
  at high altitude, 114–116
  bounding, 107–110, 112
  fast forward, 92–94, 98–103
  flapping, 70, 77, 92–122
  flocks, 116, 118–120
  formation, 116–120
  gliding, 71–83
  hovering, 103–105
  intermittent, 107–112
  migratory, 99, 112–115
  slow forward, 95–98
  soaring, 84–91
  undulating, 110–112
birds of prey, 70, 81, 84
bluff body, 14, 19, 162
boundary layer, 12, 19, 29, 49, 50, 88, 146, 147, 162
Brownian motion, 2
butterflies, 124, 128, 134, 140

camber, 15, 73, 74, 144, 145, 147, 162
Cayley, Sir George, 60
centrifugal force, 61, 85
chord, 15, 73, 162
cigarette smoke, 19, 120
circulation, 17, 21, 162
'clap–fling' mechanism, 138–140
clouds, 24, 30
  cirrus, 46
  cumulonimbus, 47
continuum hypothesis, 2

control, 21–23, 82, 88, 99, 102
convection currents (*see* updraughts)
Couette flow, 6
crystals (*see* ice crystals)
cylinder, aerodynamics of, 9, 37

d'Alembert's paradox, 21
dandelion, 10, 22, 56
density
  air, 6
  definition, 2
  hailstones, 48
  ice, 44
  water, 30, 44
dihedral, 136, 162
disseminules, 54
diving flight, 83–84, 162
downwash, 104, 117, 121, 162
drag, 9, 13–16, 21, 73, 103, 163
  form, 14, 16, 163
  induced, 15, 16
  lift-dependent, 15, 16, 75, 76, 93, 163
  normal-pressure, 13, 15, 16, 163
  parasite, 74, 122
  profile, 13, 16, 74
  skin friction, 13, 16, 75, 163
  trailing-vortex, 15, 16, 88, 118, 120, 163
  zero-lift, 16, 93
drag coefficient
  bird, 85
  circular cylinder, 40
  disc, 40
  dust seeds, 55
  hailstones, 50, 51
  ice crystals, 40
  pollen, 55
  raindrop, 28–30
  sphere, 28–30
  water droplet, 28–30
dragonflies, 128–131, 134, 140, 142

energy
  dissipation, 80, 90
  expenditure, 106, 107, 113, 150
  extraction from the wind, 83, 84, 90, 92
  kinetic, 89, 107, 120
  potential, 80, 89, 90, 107
  principal of conservation, 12
  production by muscles, 101, 105, 113, 133, 150
equation of state
  perfect gas, 3
evolution, 57, 60, 66, 68, 101, 124

feathers
  contour, 70
  primary, 70, 88
  secondary, 70
flat plate, 57
  drag, 14
  lift, 14
flight (*see* bird flight, insect flight, *etc.*)
flow field, 10–13, 25
flow visualization, 11
flying fish, 154
flying frog, 153
fossil remains, 69, 152

gliding flight
  birds, 22, 68–83
  insects, 128
  lizards, 153
  mammals, 153
  pterosaurs, 150
  winged seeds, 57–59, 67
graupel, 48
groundspeed, 84, 86, 89, 163

hailstones, 10, 37, 43, 47–52
  drag coefficient, 50
  formation, 47
  large, 47, 50, 51
  shape, 47, 48
  terminal velocity, 48, 49
haloes, 44–47
helicopter, 60
hovering, 103–105, 128–131, 136, 148, 163
  wind-assisted, 105, 136, 163
hummingbirds, 70, 101, 102, 130–105

ice crystals, 24, 34–47
  columnar, 34–37
  formation, 24, 43
  geometry, 34
  orientation, 38
  plate, 34–44
  regimes of motion, 38
  terminal velocity, 36
  unsteady behaviour, 38
ideal fluid, 21
insect flight, 23, 123–134
insects
  fast flying, 132–133, 134
  highest flying, 134
  largest, 129
  swarming flight, 136–138
instability (*see* stability considerations)

Joukowski, 21, 141

Karman vortex street, 39
kestrel, 72, 105
Kutta-Joukowski, 21–141

laminar flow, 19, 28, 49, 163
landing, 81, 95, 97, 98, 106–107, 137, 151
Leonardo da Vinci, 11
lift, 13–17, 57, 67, 72, 163
  generation, 17, 92–98, 139, 140, 142, 146, 147
lift coefficient, 16, 73, 140, 163
  maximum, 77, 95, 98, 138
  very high, 105, 138, 148
lift/drag ratio, 14, 57, 62, 79, 113, 150
lime, 10, 21, 60, 66
locust, 10, 128, 136–137

mass
  bats, 143, 145
  birds, 72, 102
  disseminules, 65, 66
  hailstones, 47
  ice crystals, 45
  pterosaurs, 150, 152
  raindrops, 30
  water droplets, 30
migration, 99, 112–115, 137
minimum drag speed, 76, 78, 79–81, 84, 88, 99
minimum power speed, 76, 78, 81, 99
moment, turning, 22, 39, 61, 163
momentum, 17, 63, 64, 104, 121, 122
moths, 124, 129
Munk's stagger theorem, 118
musculature
  birds, 70, 93, 102, 124
  insects, 124–127
  pterosaurs, 150

Newton's laws of motion, 17, 61
no-slip condition, 11

optical phenomena, 44–47
  computer simulations, 47

parachute, 56, 164
parhelia, 44
peregrine falcon, 83–84
pigeon, 10, 72, 81, 100, 102, 140
plate crystals, 34–47
pollen, 10, 54–55
potential flow, 12, 164
pressure, 10, 11, 13
  atmospheric, 4–6
  centre of, 57, 59, 162
pressure distribution, 12
  about a sphere, 25–27, 50
  on a raindrop, 25–27, 31
  on a wing, 17, 57, 72
propeller, 63, 164
pterosaurs, 149–152

quarter-chord, 57, 59, 61, 164

radar ornithology, 112, 114
raindrops (*see* water droplets)
rainfall, 10, 24–33
range, 113, 114, 164
resistance (*see* drag)
Reynolds' number, 8–10, 17–21
  aircraft, 10
  bat flight, 10
  bird flight, 10, 78, 128
  critical, 49
  hailstones, 10, 49–50
  ice crystals, 36–43
  insect flight, 10, 128, 135
  rainfall, 10, 28–31
  seed dispersal, 10, 55, 67
  sphere, 28–31, 49–50

samara, 10, 21, 59–66, 67, 164
  angular velocity of, 65
  sinking speed of, 65
scaling laws, 78, 100, 164
seed dispersal, 22, 53–67
seeds
  plumed, 22, 56
  winged, 22, 57–66
separation, 21, 28, 49, 98, 146, 164
similarity, principle of, 78, 100
sinking speed, 164 (*see also* terminal velocity)
  autorotating samaras, 65, 66
  birds in gliding flight, 79, 80, 83
  pterosaurs, in gliding flight, 150
skeleton
  bat wing, 144, 146
  bird wing, 71
  human arm, 71
  pterosaur wing, 144, 149
snow, 10, 52
soaring
  dynamic, 88–91, 151

slope, 84–85, 91
thermal, 85–88
solar pillar, 46
speed of sound, 11, 132
sphere, aerodynamics of, 27–31, 49–51
stability considerations, 21, 38, 57, 59, 61, 80–81, 150, 164
stagnation point, 21
stalling speed, 77, 78, 95, 97, 100, 164, 165
standard atmosphere, 3–5, 165
Stokes' law, 30, 55
stoop, 83–84
streamline, 12, 13, 25, 165
sublimation, 24
sweepback, 59, 165
swift, 102–103, 149

take-off, 98, 106–107, 137, 151
terminal velocity, 165 (*see also* sinking speed)
birds in stoop, 84
hailstones, 43, 48–49
ice crystals, 40–44
pollen, 55
seeds, 55–56
snow, 52
water droplets, 29–32
thermals, 54, 85 (*see also* updraughts)
thrips, 10, 135–136
thrust, 88, 92, 96, 98, 102, 165
transition, 19, 50, 165
tumbleweeds, 67
turbulent flow, 19, 40, 50, 147, 165

updraughts, 24, 54, 84, 85, 87, 133
upwash, 88, 116, 117, 165

viscosity, 5–7, 8, 17–21, 165
dynamic, 5, 11, 31, 165
kinematic, 7, 11, 31, 165
vortex, 28, 39, 116, 120–122, 139–140, 141, 166
bound, 17, 139, 140, 141, 166
horseshoe, 17, 139, 166
starting, 141, 166
trailing, 17, 75, 88, 116, 139, 141, 166
vortex ring, 120–121
vortex street, 39
vortex theory of flight, 120–122
vorticity, 17, 140, 166

wake, 12, 20, 32, 38, 40, 50, 120, 166
water
density of, 30
water droplets, 10, 24–33, 44
break-up of, 31
drag coefficient, 28–30
formation of, 24–27
size range, 25
terminal velocity, 29–32
weight, 30
weight (*see also* mass)
support, 92, 96, 104, 120
wind,
effect on dispersal of samaras, 65, 66
effect on migratory flight, 114, 137
effect on soaring flight, 83, 84–91
windmill, 63
wind-tunnel, 8, 11, 83, 124, 166
wing loading, 71, 166
bats, 145
birds, 72, 81, 84, 91, 99
pterosaurs, 150
wing span
bats, 145
birds, 72
pterosaurs, 150, 152
wing tips, 17, 88, 118, 166
wingbeat frequency, 102–103, 111, 129, 148, 149, 150

zanonia, 57–59